# IRAQ:

THE ETERNAL FIRE
1972 IRAQI OIL NATIONALIZATION
IN PERSPECTIVE

**IRAQ:**

THE ETERNAL FIRE
1972 IRAQI OIL NATIONALIZATION
IN PERSPECTIVE

# IRAQ:
## THE ETERNAL FIRE
## 1972 IRAQI OIL NATIONALIZATION IN PERSPECTIVE

**ADIL HUSSEIN**

Translated from the Arabic
by
A-W. LÚLÚA, Ph.D.

THIRD WORLD CENTRE
FOR RESEARCH AND PUBLISHING
LONDON, 1981

117 PICCADILLY LONDON W1

ISBN
0 86199 004 8


Printed in Beirut by: Express International Printing Co.

FIRST EDITION
1981

# TABLE OF CONTENTS

# PREFACE TO THE ARABIC EDITION

Since the Arab Institute for Research and Publishing has always endeavoured to provide the Arab reader with knowledge, information, research and studies connected with his life and his struggle on all levels and in all areas, we have asked the notable progressive writer and economic researcher, Mr. Adil Hussein, to evaluate an important economic and political phenomenon, which may be one of the most significant and successful battles fought by the Arab liberation movement in contemporary history, in order to explore both its obvious and hidden aspects, and to assess the reasons for its success. By doing this we hope to make it available to Arab readers and freedom-fighters both as a description of a historic event and also as a lesson in fighting great battles.

Mr. Adil Hussein has carried out his assignment with distinction, fusing the accuracy of research, the exploration of a journalist and the commitment of a thinker, without deviating for a moment from the path of objectivity.

The glorious nationalization battle fought by the Iraqi people, supported by the entire Arab nation, under the leadership of the Arab Ba'th Socialist Party, is not an isolated phenomenon. It is linked with the nature of the party's national objectives on liberation, and in the course of its struggle, reverberates with the aspirations and interests of the people. It was also connected with mottoes raised by the Party at an early stage: "Arab Oil for the Arabs". It was endorsed by a decision emphasized by the Political Report of the Eighth Regional Congress of the Arab Ba'th Socialist Party in Iraq. That decision reads: "Economic independence is considered by the Revolution as a central aim of liberation, parallel with, and complementary to, political independence. Without real economic independence, political independence loses one of its basic means of support, and is rendered always vulnerable."

We may not deviate from the truth when we say that the events which followed nationalization proved that the Party leadership was able to put into real practice, and in a courageous and pioneering manner, the motto of "Oil in the service of the battle"; and that was on the eve of the October War of 1973.

The author has spared no effort to study all the aspects of the story of nationalization and the oil policy in Iraq, by his lengthy sojourn in Iraq, his interviews with responsible people on various levels, his exploring of minute details, his visit to the oil-fields and his scrutiny of numerous documents on this question. The reader will note the author's familiarity with the oil industry and with the history of the monopolizing companies, their methods, exploitations and conspiracies. In recognizing this effort, we have also to recognize Mr. Adil Hussein's ability and success in ascertaining a desire expressed by President Saddam Hussein who said:

> "The study must be concerned not merely with the idea of nationalization... or merely with the idea of the liberation of oil riches but with the political attitudes and actions responsible for the ripening and crystallization of the idea, and the implementation of it into actual practice."

The Publisher

# INTRODUCTION

*America Loots Arab Oil* was my first book (1957). In the final chapter I wrote:

"Yes, nationalization. Oil belongs to the Arabs; and to the Arabs alone. We will not accept tutelage, and will re-claim our looted oil-rights. This is the main lesson to be learned from this book, and the main aim which we must fulfil. But we have seen from this survey that this right is impeded by numerous obstacles; and this is normal. Can we imagine a gangster, threatening people's livelihood, attacking whatever is dearest to them, and subjecting the area to his authority, if he were not fully armed with weapons, in the use of which he is well trained? Of course not!

"This is also the case with the oil-gangs; they are not different from the gangs which they excel in presenting in their films. But the oil-gang is more dangerous and ferocious; it is a gang extending its authority east and west, coming to our country on board a Sixth Fleet, with missiles and atomic weapons, with cries and warnings and manoeuvres, not mounted on a horse and brandishing a silly gun. This oil-gang comes to our country and seizes the oil by usurpation.

"And we here, in our country, know that this gang is strong, and has a strong Sixth Fleet. But we also know, and have an unshakeable belief that we all can, and are willing and able to get them out of our country, and regain our oil from their hands...

"Dear Arab reader... it is within our power to conquer this stubborn enemy [America]; it is no longer the era of pirates, and, in spite of their armies, development and history have given our Arab people the wherewithal to conquer our enemies.

"The defeat of our enemies shall be achieved, and it could not be achieved except by one weapon, and that is unity... Let us be united, then we shall put an end to the enemy.

"Let us unite our ranks in every Arab country, and let us cleanse these ranks from traitors and imperialist agents. Let us unite the national ranks, the enemies of imperialism, and let us bring them together in a composite national front. This front is the first fortress from which we shall attack the enemy, and at which he shall aim his fire.[1]

When I read these words, now that twenty years have passed, I remember my youthful emotion. But the emotive expressions do not only reflect the enthusiasm of a young man who, several months earlier, was carrying his arms to fight aggression on the Suez Canal. In fact the emotive expressions reflect, in addition, the warm temperature of the entire Arab area. The Arab masses were filled with self-confidence after the victory in the first major battle of nationalization. The masses were, like our generation, resisting the Baghdad Pact, and blocking the way before the Eisenhower Doctrine attempted to fill the 'vacuum' with an American existence. Great were our expectations!

The success of the nationalization of the Suez Canal, and in the checking of the Triple Aggression, led us on in our imagination to the oil battle, which would put an end to imperialist influence, and point our Arab nation towards new horizons of development. I am now of the opinion that the extreme nature of my exhortations, at the conclusion of my book, was a reflection of the climate of opinion then, rather than simply a portrayal of my emotions as a youth.

In any case, I can well remember that I wrote the book about the serious significance of oil, the nature of relations controlling its production, and about the development of the struggle inside the international cartel – and my eye was on the possibility of revolution in Iraq. This is not to say that I was endowed with special prophetic powers; this hope was common among the revolutionary youth in Egypt. And, in spite of all the suppression exercised by the royal authorities, we were expecting an explosion in Baghdad. From Baghdad the oil nationalization battle was to begin.

It was only a few months later that the revolution broke out, in July 1958. The first edition of my book was out of print by then. Dar Al Fikr brought out a second edition, most of which was sent to revolutionary Iraq.

Those were beautiful days when excitement reached a climax. But, we had forgotten to unite the ranks of the nationals, enemies of imperialism, and to bring them together in a composite national front. We had forgotten that this front is the first fortress from which we should attack the enemy, and at which the enemy was to aim his fire. We forgot this principle in Egypt and Iraq – and in other Arab countries, so the initiative began to ebb.

Much water has flowed under the bridge since then. That period seems remote now, and numerous events, uprisings and setbacks have occurred. In the meantime, what has happened to the question of oil? Has Arab oil been restored to the Arabs? Undoubtedly, the Arab world has taken a step towards the liberation of its oil wealth. Progress in this direction has not been at the same tempo in every country. The oil country which nurtured our hopes in 1958 did not disappoint us. After a tiring and bloody course, Iraq occupied the position of the pioneer.

It is important to underline two facts. The first is that a realization of the extent of the struggle in the Arab world cannot be separated from an understanding of the oil question. An Arab, or a politician, who does not fully understand the oil question will undoubtedly fall into gross error. Speaking about oil, naturally, does not mean here the technical details about production transport, industry and usage of oil. As an Arab nation, what concerns us in the first place is to follow up the international blocs. The duty of the Arab nation is to face these strategies with a counter Arab strategy, able to deal with the many variables involved in the battle.

This leads us to the second fact: the Arab oil struggle has now gone past the stage of simple mottoes about nationalization and national control; actual practice, since the Fifties, has deepened and formulated our concepts. No sound Arab strategy can be drawn up without these defined concepts and acquired experiences.

Oil strategy and defined modes of action cannot be the invention of a few sharp minds; they cannot be formulated apart from direct experience in the struggles of recent years, especially the experience of the Iraqi revolution. Those who led the revolution of 17-30 July 1968 in Iraq closely studied, the former experiences in Iraq and the Arab world, and emerged with a radical line, correcting numerous deficiencies in the progressive experiences in the area during the Sixties. The Revolutionary Command handled the oil question in the same manner, understanding the experiences of the former struggle, defining the theoretical framework of movement in this field, then adding depth and modification, by virtue of actual practice. Now it is imperative to crystallize the results, in preparation for present and future battles.

Tracing the achievement of the Iraqi revolution in all fields, through the leadership of the Arab Ba'th Socialist Party, is a matter of theoretical and practical importance not limited only to Iraq. It is of vital importance to all Arab revolutionary forces, and even to those outside the Arab world. If a definite understanding of the complex oil question is urgently needed in this area, then Iraq is the basic school where we can learn the principles concerned. Iraq is the only country which can exercise full sovereignty over its oil. Iraq is also capable of planning and exercising an independent oil strategy, reflecting totally national interests, and reflecting also the interests of the oil-producing countries. This was the incentive which led me to explore what has been achieved in this country.

Among the consequences of successful nationalization is that an independent and comprehensive oil policy emerged in a country which is objectively capable of influencing the surrounding area. Iraq is a country of standing in the Arab world, with a large number of oil technicians and a long and uninterrupted history of national struggle. In addition, Iraq is not a "pigmy" state but enjoys a reasonably sized population as well as various natural resources. The growth and development of Iraq took place prior to the emergence of its oil. the ability of Iraq to diversify its income resources is consequently easier than with some other Arab oil countries.

As an oil state, Iraq has links with the Arab countries, which together produce 45% of world oil (excluding the United States and the socialist bloc). This adds to the authority of Iraq and to the extent of its influence. What is even more important is that Iraq enjoys a particularly influential position with regard to oil reserves, and in the relations between consumer and producer states, and also in the relations inside OPEC (the Organization of Petroleum Exporting Countries). The owner of large reserves is like the owner of large shares in a company — his voice is louder than others and his word has to be respected.

Iraq is the only country that was not included in the campaign in the Sixties instigated by the oil companies for the search and exploration for oil. This campaign extended everywhere, using all available geological knowledge, and equipped with modern theories and equipment. Iraq was not included in this campaign because of its struggle with the companies in that decade, which once again points to Iraq, above all Middle East producers, as offering the best prospects for future development. The results of modern research in Iraq have not been published, and the official estimates of the proven reserves are surrounded with strict confidentiality by the Iraqi authorities. Available Western estimates are based on analyses (by American experts) of the results of a survey undertaken by the Iraq Petroleum Company (IPC). According to these estimates, the reserves of the southern area of Iraq exceed 80 billion barrels. If we add to this the 15 billion barrels that are the proven reserves of the northern area, then Iraqi reserves are second only to Saudi Arabia's.[2]

The technical consultant to the Iraqi National Oil Company (INOC), announced even more optimistic estimates in July 1972. He said that the concessionary companies had estimated the proven reserves of the fields exploited by them at 32 billion barrels. This does not include the reserves owned by INOC which some independent American sources estimate at 75 billion barrels. Iraqi geologists think that this figure may double in the next few years, in the light of the new discoveries west of Qurna and along the north Rumaila field, and also in the light of INOC's ten-year plan. If this is so, then Iraqi reserves will jump to more than 180 billion barrels, which means according to the technical consultant that Iraqi reserves may exceed the proven reserves anywhere in the world.[3] (Saudi proven reserves by the end of 1973 were 148 billion barrels.)

The fluctuations in these estimates do not concern us at the moment, since the latest figures have not been released, as we have said. What concerns us more generally, is that Iraq is in a position to influence the future of world oil.

We are therefore looking at a country playing a pioneering and progressive political role. What doubles the importance of this role is the importance of the area in which Iraq exercises an influence. Moreover, Iraq uses its oil independently, and in accordance with a strategy based upon its advanced ideas.

In spite of all this, we find that Western writing about oil policies ignores the present and potential role of Iraq. It hardly deals with what happens in the area, except in a curt manner, betraying a deep rancour. The international cartel companies – backed by imperialist countries – have not stopped and will not stop conspiring against the path followed by Iraq.

Imposing a news ban is one way to encourage secret conspiracies – a well-tried method with regard to revolutionary and socialist experiments. Progressive forces have to do the opposite. We have to direct the greatest amount of light on what has been achieved, because this encourages the resolution to defend the victories which form an asset to all the Arab masses. A follow-up of these victories revives the hopes at a time when imperialism is accelerating its attempts to regain control in the area.

I hope that this book will contribute towards a clarification of the basic aspects of Iraqi oil policy, within the general framework of the revolution. I have deliberately refrained from treating the subject in a fully academic manner; yet I have committed myself to following scientific methods concerning the data which form the basis of the study. Despite approval of the Iraqi oil experiment as a pioneering and a comprehensive experiment, I still believe in the objectivity of the book, because this judgment was not the starting point of the study, but its result.

## NOTES

1. A. Hussein, *America Loots Arab Oil* (Cairo, 1957) pp.86-90 (Arabic).
2. M. Field, *A Hundred Million Dollars a Day* (London, 1975) p.48.
3. S. Shareef, *Arab Oil Reserves and their Role in the Future of the World Oil Industry,* INOC publications (n.d.).

CHAPTER 1

# THE DIFFICULT DECISION

In the Kirkuk fields, I visited a site displaying a sign saying:
Baba Gurgur – the first productive well in Iraq.
First Depth: 1,521 feet (14 October, 1927).
Second Depth: 2,000 feet (1933).
Closed: 28 July 1940.
Duration of Drilling: 5 months.

At a second site, there was another sign of a similar size, giving the following information about another oil-well, the well of "Peace and Solidarity":
Drilling started on 20 July 1972.
Duration of Drilling: 14 days.
Depth: 1,670 feet.
Production: 3 million tons per year.

Our car made the distance between the two sites in less than five minutes; but one's mind cannot help thinking of the long and bloody course of the Iraqi struggle to move from the first well to the second. What "Baba Gurgur" stands for is the exact opposite of "Peace and Solidarity". The contradiction is underlined by the fact that the first well was drilled in 1927 and the second in 1972.

Baba Gurgur was not only the first oil well in Iraq, it was the first productive well in the Arab oil-exporting countries.[1]

The discovery came as a complete surprise. Reports reaching London in 1927 spoke of a number of unsuccessful drilling attempts in the Kirkuk area. In the midst of acute frustration, an emergency meeting was held by the IPC Board of Directors. Some were complaining of high costs without the promise of results. The majority seemed to favour a cut-back in labour and expenditure, and a postponement of any new drilling projects. But a minority strongly disagreed and at their insistence the work continued; the site, however, had to be defined.

The chief company geologist explained that there was a slim chance of finding oil in Baba Gurgur. But the representatives of the American companies in IPC had sent their expert geologist Mr. Trowbridge to the area. His report was enthusiastic about the possibilities of oil at Baba Gurgur. The American opinion prevailed in the end, and Baba Gurgur was chosen, without enthusiasm, by the majority of members present.

In Egypt, the Ghardaga field was discovered in December 1913, so this may be considered the first Arab field. But since we are concerned with the oil-exporting countries, and with the main production areas, Baba Gurgur remains the first well.

Most of the drilling teams, the equipment and the administrative headquarters were in the vicinity. The decision was taken to begin drilling at Baba Gurgur site, emphasizing that it was to be the one and only attempt.

Drilling actually started on 30 June 1927, and there was little interest in the progress. With the passing of days and months, the disappointment grew to the point where a new order to stop was imminent. By midnight,14 October, the drilling had gone beyond 1,500 feet, and the man in charge of the rig was determined to continue. The rotary drilling was replaced by turbine drilling, and he ordered the Iraqi workmen to remove the casing, leaving 500 feet of mud at the bottom of the well, as a buffer to check back the oil – if indeed it was there. These were merely precautionary measures.

The turbine drilling started. The man in charge of the rig stood with his two colleagues (all three were American) beside the steam engine which was pushing the drilling-bit into the rocks. In the calm of the cool night the conversation was about the eternal fire whose light was dispersing the surrounding darkness.[2] The site of the fire was rather ironic. It seemed such an obvious indication of the existence of oil in that area.

Drilling continued for a while. The man in charge ordered that the drill be drawn up in order to clean off fractured stone from the well. The men began to carry out the order.

And here was the surprise.

A faint noise began to come from the depths. Then gradually the noise developed into a roar, followed by a spurt of mud and oil which grew into an upsurge, inundating the base of the drilling tower.

It was a complete surprise, and no safety measures had been taken. The following morning witnessed a column of oil like a giant cypress tree swaying in the wind. The height of the column had reached 200 feet and was still increasing. The oil was spraying people for miles around. The camp was flooded with up to half a metre of oil, which was pushing its way to the valley, and even into the Tigris river.

It was not until nine days later that the surprise was finally placed under control, at 3 pm, 23 October 1927.[3] The story of the Baba

Gurgur discovery became an interesting tale in the history of oil operations. But this well was an important turning-point not only in the modern history of Iraq, but also in the history of the Arab world and the Middle East. Iraq had become a centre of oil production, with all that this signifies in the way of struggle, conspiracies, pressures, and even wars.

The well of "Peace and Solidarity" was not drilled in an atmosphere of indifference. The decision was not taken casually, nor did the people follow the news of the drilling without interest. On the contrary, every one followed events with confidence and passion. The drilling itself entailed much self-sacrifice. Once the preliminary studies were complete, the drilling was undertaken with more modern equipment and better experience, so that when the oil gushed out it was under control from the first moment. The surprise was that drilling took only fourteen days, while it was expected – by prevailing standards – that a drilling of that depth should have taken a whole month. But this surprise – in the terrible heat of July – was not a technical one, but rather a human surprise, reflecting the will-power of the men at the rig. It is relevant to note that this time they were all Iraqis. And though it was not new that an Iraqi staff should take full responsibility for a rig, it was new that they were extracting oil from that very Kirkuk field, in accordance with plans and instructions coming from a national Iraqi administration. They were developing oil production in Kirkuk in order that the resultant wealth should return to the Iraqi people. That was the new and important factor.

The drilling of Baba Gurgur was a dramatic story. Yet the epic of "Peace and Solidarity" did not lie in the struggles and ironies of drilling, but in the entire revolutionary operation of nationalization. And even when foreign faces were to be seen among the Iraqis, they were of a different kind from those who surrounded the Baba Gurgur well at the celebration of its opening. They had come this time from various parts of the world as members of the Peace and Solidarity Congress, to support the Iraqi people in reclaiming their oil wealth.

Reaching the stage of the well of "Peace and Solidarity" was no easy matter. The nationalization of oil was at the top of the list of objectives for which the Arab Ba'th Socialist Party was struggling. Since the July revolution of 1958, it had seemed more than once that this aim must be put on the agenda. Yet, when the circumstances were favourable in 1972, and a decisive step was imperative, hesitation prevailed. This is natural, as there is a wide gap between theoretical conviction and practical application. The one is much easier than the other. How easy it is for a man to believe in the necessity of revolution, for instance, and how difficult it is to make revolution a reality.

Four years after the nationalization decision of 1 June 1972, Iraq had left behind all the psychological tensions which preceded and accompanied that decision. "Talking about nationalization became familiar," to quote Saddam Hussein, "because the experiment was carried out... and was successful; and because the one who jumped into the river was able to swim skilfully and to reach the other bank safely."[4] But, when we recall challenges which then seemed insuperable, when we remember the fears which prevailed, we realize the extent of revolutionary imagination and courage which were needed to take the decision. It is enough to recall some of the highlights in order to re-create the general picture.

The first move came from the headquarters of the National Council in Baghdad, in the room adjacent to the office of Saddam Hussein, the headquarters of the staff in the oil battle. It was the midnight of 31 May 1972. Present were a number of distinguished national personalities. The question posed was very simple: What's to be done?

Saddam Hussein briefed those present on the latest developments in the oil battle. The evening of that day witnessed the last discussions held with the delegation of the oil companies. In that ninth session, the head of the Iraqi delegation said: "We categorically refuse to accept the offer made by the companies. There is no alternative response now or in the future." And he added, "We demand that the companies should present a more acceptable offer, and the deadline is 11 am tomorrow morning. This is flexibility on our part."

Saddam Hussein did not hide his opinion that the companies would not offer anything new and that the negotiations had reached a dead end. The last chance offered to the companies to adjust their position was not enongh for them, and in any case they were not ready to offer any new concessions. Mr Stockwell (IPC Managing Director and chief negotiator) declared: "The meeting tomorrow morning, though I am interested in it, will not be, as far as we are concerned, free from the points which the head of the [Iraqi] delegation considered provocative".

There was nothing new, then, to be expected, even if another meeting were held on the morning of 1 June. What was to be done?

Those present at the conference had to present their opinions and recommendations. They were all of the opinion that oil nationalization was a strategic aim. The question had been decided in principle quite some time ago. Even under the circumstances of 1966, a number of patriotic personalities submitted a memorandum to the Prime Minister, emphasizing that "the oil sector is of the highest importance, not only in the national economy, but in the long-term political strategy, which makes it necessary to deal with the oil companies working in Iraq in accordance with a long-range plan – political, economic, military, internal, Arab, and international – in order to nationalize the oil at a suitable time"[5].

The idea of nationalization was thus not far from the minds of the patriotic and progressive leaders. But at this meeting and after an extensive discussion of various possibilities, all those at the conference expressed their reservations about the complete nationalization of IPC. They were of the opinion that the government should resort to extracting some of its rights through unilateral legislation, but that the legislation should not amount to complete nationalization. The most radical suggestion called for the nationalization of 51 % of the IPC shares.

This decision does not reflect badly on the patriotism and loyalty of the majority of those who attended this meeting. It rather reflects the seriousness of the moment when a critical decision has to be made. As for Saddam Hussein, he sat listening for most of the time, not interfering except in discussions of detail, although the statement on comprehensive nationalization had been ready at the Command Headquarters for quite some time, ready to be broadcast on the first of June. In fact the form of nationalization had been discussed in detail and with great concentration since January and February 1972, and the two-week ultimatum formed the period of final preparation for implementation through "The Follow-up Committee on Oil Affairs and Implementation of Agreements" which was headed by Saddam Hussein.

The second move took place a few months before this session, when a number of patriotic economists and technicians were asked to give an opinion about the possibilities of nationalization. All their reports came to the conclusion that nationalization was impossible. The problems which had killed the nationalization experiment in Iran were still applicable to Iraq. All their calculations stressed that the national staff was incapable of planning for field development, or project design, and that the question needed years of experience, large financial resources, and a large scale of international co-operation, all of which were only available to the foreign companies. The reports and economic studies also emphasized that national experience in marketing was very limited and, consequently, the Iraqi administration would fail in marketing even a minimum quantity of oil. So Iraq would not get even what the companies were paying at the time, which would lead the country into financial straits for the imperialist and reactionary forces to exploit, thus leading to the downfall of the system.

The reports did not overlook the fact that the structure of the Iraqi economy was more deformed than the Iranian economy at the time of Musaddaq. Oil revenue formed 12 % of the Iranian budget, while oil revenue in Iraq formed 52 % of the income of the regular budget (1971-1972), and 88 % of projected investments (1971) and, generally speaking, these revenues formed 46 % of the gross national product (1972). These figures showed that the risk was very great, and that the economic catastrophe would be considerable if Iraq should fail in marketing its nationalized oil.

The Command of the Arab Ba'th Socialist Party (ABSP) sought constant technical advice at its meetings before taking decisions. The topics under discussion were studied by specialist committees, which submitted their reports to the Command. Sometimes the Command requested the presence of the committee spokesman at the meeting, in order to explain and discuss, then he would withdraw, and the Command would continue the discussion of the subject, and finally reach a decision. But in the question of oil particularly, and because of the reservations apparent in the previously mentioned reports, amounting sometimes to complete horror, all subsequent studies were limited to the political Command.

The third move is reported by Mr Tariq Aziz[6] MRCC, Deputy Prime Minister Mr Aziz was at that period the editor of *Al-Thawra* daily, the official organ of the Ba'th Party. By virtue of his position, he was close to the pulse of the follow-up committee, and this was reflected in his daily directions to the editorial team. In fact the tone of *Al-Thawra* in those days, and particularly since the ultimatum of 17 May, indicated that the confrontation was imminent. On 31 May, for instance, the leading article read: "Tomorrow the monopolies shall know for sure that we mean every word we say." On the first of June (the deadline) the headline read: "The first of June, the day of the People's victory over the monopolizing companies". Despite this atmosphere, says Mr Aziz, there was no Ba'thist or progressive writer among the editorial staff who dared utter the word "nationalization". They were expecting unilateral measures, but they never imagined that the issue could be settled at one radical blow. Even when President Ahmad Hassan Al-Bakr announced in the evening of 1 June that "the Revolutionary Command Council has decreed, in the name of the People, the issuing of the Law No. 69 of 1972, nationalizing the operations of IPC, which shall be valid from this first day of June" – even after this declaration, some people were hesitant, and came to Mr Aziz to ask what was meant by the nationalization of IPC operations. "They imagined that the expression must have another hidden meaning. I was obliged to assure them that the decision was not open to interpretation, and that it meant nothing less than comprehensive nationalization."

This was the case with educated and well-informed people. And the same phenomenon was encountered at the popular level. During my visit to the nationalized company in Kirkuk, I asked scores of workmen and technicians whether they had expected the decision announcing comprehensive nationalization on 1 June. They answered unanimously that the decision came as a surprise to them. They had followed the course of the struggle with IPC, and the decisions taken and the level of political recruitment indicated that something was afoot but they did not expect it to be full nationalization.

The fourth move came from outside Iraq, from the enemy side. This was a report by the French News Agency, broadcast on 30 May 1972, the day when Mr. Stockwell of IPC left London for Baghdad. The report alluded to the IPC calculations when presenting their final suggestions at the negotiations. They suggested that the Iraqi side review "the facts" before taking any decisions. The report read:

> "The companies are speaking in fact from a strong position; much oil at present is available in the world; so the companies can easily do without Iraqi oil, which can hardly represent 3% of the world oil production, let alone the fact that the size of the companies' investment in Iraq decreased a long time ago. The Iraqi government may nationalize the possessions of these companies, but if the government could, as is probable, continue the production of oil directly, or if need be, by resorting to the Soviet experience in this field, then the appropriation of the companies' assets will never solve the problem of oil marketing. The Iraqi government, deprived of the international distribution network of the companies, would not be able to dispose of 30-50 million tons of oil a year under the

present circumstances, except by reducing prices at the expense of income.

"The companies may resort to the law to prevent purchase of oil imported in this manner, by confiscating it, a measure which has proved its validity several times in the past. Iraq may export some of its production to the Communist countries. But these countries have a limited intake capacity since they are actually buying INOC products from an oilfield which the IPC is claiming [ (i.e.) North Rumaila Field]"

The IPC calculations, then, were based on the assumption that Iraq's attempt at nationalization was impossible. At the same time, the company was baring its teeth and indicating that its weapons might be used if the Iraqi decision-maker were to lose his senses.

Now we come to the fifth move, again from the outside world, this time from the friendly side: the Soviet Union. The companies and the Western press have always linked the Iraqi decision with a preceding agreement with the Soviet Union. The *Nouvel Observateur* wrote (on 18 June 1972) and with the confidence of an insider: "Nationalization was effected with encouragement from Moscow leaders, or with their consent at least, while they were closest to the Baghdad leaders". The *Washington Post* had written with the same confidence (7 June 1972), "There is no doubt that Iraqi behaviour is backed by Soviet encouragement".

This is the claim circulated in the West. But what is the truth? It is that the 17-30 July revolution was careful, from the start, to buttress her relations with the socialist countries, and the Soviet Union in particular. The Revolutionary Command realized, through objective calculations, that buttressing these relations was a necessity.

"The alliance with international powers enjoying political, military, and economic strength and being close to us in principles, objectives and interests, in the process of our national struggle with the enemies (international imperialism and Zionism with its presence in our occupied land), is not only a sound process, but also a highly essential one, provided we preserve the distinguishing characteristics of our Arab revolution, its independence and free will... The Soviet Union and the socialist countries, despite differences or lack of agreement between us and them on a number of questions, are the closest advanced and strong countries to us, in principles, objectives, and interests."[7]

In accordance with this programme, the Iraqi command accepted assistance from the Soviets, as well as from the socialist countries, in the field of direct national production. Their technical and financial assistance was an important support in the struggle towards the liberation of oil wealth. Again, in accordance with the same programme it was natural that the Soviet Union should be in the picture, with every increase in confrontation with the monopolizing companies. In the first visit to the Soviet Union by Saddam Hussein (August 1970), exchange of opinions began and communications continued after that, especially during the second visit by Saddam Hussein (February 1972). In the light of all this, the first article of the Iraqi-Soviet Treaty (19 April 1972) provides for the Soviet Union's support of Iraq in the course of the liberation of its oil wealth.

Naturally, the Soviet attitude was not against the Iraqi right to nationalize as a final aim. However, it advised that confrontation should advance by stages, to try and achieve an alternative solution to comprehensive nationalization. The Soviet Union tried, by various means, to frankly acquaint the Iraqi Command with the Soviet Union's real ability to support such a battle. The European socialist countries, headed by the Soviet Union, would be unable to take in Iraqi oil in the event of its marketing efforts failing. At the same time, their capacity to give financial aid could not compensate Iraq for all that had previously been gained from the companies, especially if it became necessary for most of this to be paid in hard currency.

It seems that the Soviet Union had doubts about the ability of the Arab Command to investigate thoroughly before taking large decisions. These doubts were born from earlier bitter experience. The Iraqi Command tried to dispel these fears, and explained their estimations that the nationalization decision at worst would not cause undue danger or trouble for the Soviet Union. Iraq would not demand from the Soviet Union more than a friend would ask, and the limit of the Soviet Union's capacities was appreciated by the Iraqi Command. The issue belonged to the Iraqi people, and the people alone should be asked to offer the sacrifices demanded by the struggle.

Yet the Soviet side still had its reservations. It is true that Tass news agency published a statement supporting "the decisive measures which the Iraqi government was forced to take" on the day following the nationalization, a statement which was of important moral support to the Iraqi masses, yet the political Command realized that what was needed was more than moral support. The Command also realized that if material support was not secured before the decision of nationalization, it was necessary to continue the effort after that. The Iraqi Foreign Minister was supposed to go to the Soviet Union before 1 June, but, unfortunately, the visit was not possible before 7 June. After the talks with Alexei Kosygin "the Soviet side declared, in a joint statement, their support of the measures taken by the Iraqi government, and of the just struggle of the Iraqi people, aiming to reclaim its legitimate rights to its oil wealth and the use of oil resources for the development of the national economy and the welfare of the people." This time the statement had more than a moral value; the Soviets had become convinced that the Iraqi Command had not entered the battle without lengthy preparations.

To record these facts does not belittle the value of the political, economic and technical support afforded by the Soviet Union in the Iraqi oil battle, before and after nationalization. I have noted these facts to emphasize certain points.

Firstly the nationalization decision was one hundred per cent an expression of the revolutionary will of the Iraqi Command. That Command alone took the responsibility for any possible dangers.

Secondly the Iraqi Command proved that it had an integrity of principle and a degree of political experience in keeping complete silence about this disagreement. Even when matters were settled, the Party Command did not fall into the trap of imperialist provocations about Moscow's decision. Not a single responsible person made any comment. There was complete vigilance lest small differences should develop into news battles harmful to strategic interests.

And thirdly the Command proved again that the decision-maker who takes responsibility and accepts a calculated risk, in order to make a revolutionary leap forward, is able to render the dialogue with the adversaries fruitful, thus making the allies renew their calculations.

The above-mentioned moves form five aspects of a single picture. This shows that all sides, irrespective of their various intentions, had calculated that nationalization was impossible. Undoubtedly, this picture suggests two immediate questions:

Firstly, why were these parties mistaken in their evaluation? And, secondly, was the nationalization decision, in the face of opposition and reservation, a leap into the dark? Or was it an uncalculated risk which met with good luck?

NOTES

1. In Egypt, the Ghardaqa field was discovered in December 1913 so this may be considered the first Arab field. But since we are concerned with the oil-exporting countries, and with the main production areas, Baba Gurgur remains the first well.

2. In Baba Gurgur, particularly, there has been a fire burning from time immemorial. The oil experts considered this fire as a sure indication of the existence of oil. The eternal fire is a result of natural gases penetrating up through the soil. Oil itself was abundant at that site.

3. *Vide* L. Mosley, *Power Play* (Weidenfeld & Nicholson, London 1973).
4. Saddam Hussein, "On the Nationalization Battle", June 1, 1973.
5. M.S. Hassan, *On the Nationalization of Iraqi Oil* (Beirut, 1967), also I. Allawi, *Iraqi Oil and National Liberation* (Beirut, 1967) p.153 (Arabic).
6. T. Aziz, conversation with the author.
7. *Political Report of the Eighth Regional Congress*, January, 1974, Part Nine. (Arabic).

## CHAPTER 2

# ONE: OIL – CARTEL – AMERICA

**One:**

To answer the first question we have to resort to a flashback – a quick survey of facts and information concerning the situation up to 1972 which were present in the minds of those who thought of nationalization.

Firstly, oil is an unusual product, inseparably linked, particularly in the last quarter of the century, with the spirit of the age. Without it, the achievements of modern civilization would be impossible. and this product is not distributed with "equality and justice" in all countries and areas, since the producing areas are limited, while the consuming countries are spread all over the world. Thus the value of crude oil carried by tankers from producing areas to consumers' markets cannot be compared with the value of any other product in international trade.

Secondly, as a consequence, the oil industry developed fast and became, by all criteria, the greatest of all industries in size, and the only industry which concerns every country in the world, without exception.

And thirdly, whoever controls this industry possesses huge power. Consequently, oil history is the history of struggle among the big powers competing for areas of influence. Each power endeavours to secure its needs, and some try through monopolizing the "secret of life" to control others. Oil history is black. It is full of wars, conspiracies, coups d'état, assassinations, looting and bribery... The United States – since World War II – has endeavoured to strengthen her role of controlling the world by securing control of oilfields outside her territories.

I shall now present a brief survey of the basic background, particulary from World War II up to the time of the Iraqi nationalization of oil.

We need not talk extensively about the importance of energy in an industrial state.[1] It is enough to note from the start that, during the period in question, the rate of increase in energy consumption in the world was not less than 5% per year, and the rate of increase in oil production in particular was more than any other rate of increase in economic affairs. In the period from 1945 to 1950, the amount of oil produced doubled to reach 500 million tons. The world needed ten years to reach another 500 million tons, thus reaching 1,000 million tons in 1960. The third 500 million tons did not take more than five years to achieve, and the fourth 500 million tons took only three years, when total annual production stood at 2,000 million tons. The present expectation that annual production will reach 3,000 million tons in 1980, and about 4,000 million tons in the late Eighties, sounds quite realistic.[2]

This dramatic increase called for a drastic change in the structure of energy used. In the countries outside the socialist bloc, oil became the first source of energy. This development, as can easily be recognized, led to political results of great significance. The struggle for oil inevitably became more acute.

A comparison of the proportion represented by oil and natural gas in the sum total of energy used in the years 1950-1960-1970 will give us this picture.

In western Europe, the increase was 10% - 36.5% - 68%

In Japan, the proportion in 1950 did not exceed 7%. But in 1960 the figure jumped to 40.5%, and in 1970 it reached about 75.5%.

Expectations for the structure of energy to be consumed in 1980, according to 1972 estimates, give the figures for oil and gas in North America as 78%, western Europe as 82% and Japan as 84%[3].

It is noticeable that the US was the first country to use oil as a primary and basic source of energy, then western Europe and Japan developed the same energy structure.

European socialist countries are following the same trend, though coal has until recently formed a high proportion of energy resources. In 1961, oil and natural gas formed 30.2% of the energy consumed and coal 68.9%. In 1974, oil and gas went up to 51.4 and the coal down to 44.9%[4]. The same tendency is traceable in the so-called developing countries.

This demonstrates a general tendency, then, or a law connected with technological development in the contemporary world, and with the comparative efficiency of available sources of energy. Oil is not expected to abdicate its position on the throne of energy or, as a consequence, of politics, whether in the near future, or even by the end of the century.

But the oil which is required in increasing quantities by all countries and continents is found in a limited number of producing fields, with various amounts coming from various fields. In 1970, for example,

North America produced 605.4 million tons a year; the Caribbean area, 202.7 million tons; South America, 58.3 million tons; western Europe, 15.3 million tons; the Middle East, 829.1 million tons; Africa 259.5, million tons; the Far East, 77.6 million tons; the socialist countries and China 433.5 million tons.[5]

In each of these areas, each country had a different share of the product. The first ten countries in 1975 were graded according to their resources as follows:

The Soviet Union, the United States of America, Saudi Arabia, Iran, Venezuela, Iraq, Kuwait, Nigeria, Canada and Libya. But this order is constantly changing. The US, for instance, came before the Soviet Union until 1974. Venezuela came before Saudi Arabia and Iran until 1970. Kuwait was before Iraq until 1974, and Nigeria was a new arrival into the market a few years ago.

Yet we can say that the three main points of oil production, irrespective of constant change in their position in the list, are: the Soviet Union, the US and the OPEC group.

The US and the Soviet Union have adopted a policy depending on self sufficiency. In spite of the relatively high cost of American oil production, the policy is to limit oil imports (not to exceed 10% of local needs) and the American oil industry has become a system independent of what is happening in the rest of the world. But, in spite of the intention of the US declared policy to achieve self-sufficiency in energy sources, the US has become one of the deficit states (i.e. its oil imports exceed its oil exports). In 1948 the deficit was rather limited and the US tried to keep it down to a low level. With the beginning of the Seventies studies were initiated as a result of an emergency energy shortage.

Every effort is now made to decrease dependence on imported oil (which does not sound realistic in the short term); and this was the incentive behind the Energy Independence Programme of 1980.

The important point is that American oil does not come into international trade, and it no longer supplies the needs of the large local market. The situation is different in the Soviet Union; with the discovery of new oil areas after World War II (Baku 2 and Baku 3) the replacement of coal by oil and gas proceeded rapidly. In 1973, oil and gas together accounted for more than 60% of energy used, and the rate is expected to go up to 75% in 1980. In spite of this, the Soviet Union is able to maintain self-sufficiency, and may insist on this for political reasons, though it may not be economically useful for the remote parts of the Soviet Union. In addition to this, the Soviet Union is able to export oil not only to socialist Europe but to western Europe as well. These exports went up from three million tons in 1955 to 40 million in 1969.

Yet the largest part of Soviet production is used for local consumption, and the amount going into international trade is still limited. This means that the area which possesses a surplus which the majority of the consuming world can rely upon is the third area of oil production – the OPEC countries. These countries now represent 90% of production outside North America and the socialist bloc and 95% of the total amount of oil in international trade. The Middle East and the Arab countries represent the two basic areas inside this third area of oil production.

Pierre Dumas wrote: "What inauspicious times these are which make the oil pour out of the Arabian deserts, while it is hardly traceable in our woods and valleys, where thrift and stability would help explorers and engineers move with facility and ease? "[6]

Whether the times are inauspicious or otherwise, this remains the case.

Now, it has not been simply its mere natural and chemical qualitities that have caused oil to take the lead in the the world of energy. Without discovering enough reserves, and without securing required provisions in a regular manner, and at "reasonable" costs, it would have been impossible for industrial states to increase their reliance on oil as much as they did. In addition to these considerations, western Europe and Japan do not possess oil-rich fields in their territories, and the US need for imports is growing. Therefore, there is a steady flow of oil to the big consumers from the producing countries. This demands efficient and stable economic establishments, which in turn demands political manoeuvering to regulate the activities of these strategic establishments inside the producing areas, as well as inside the consuming markets, so as to ensure the economic interests of Western allied countries, and within the framework of the general strategy of these countries and the US, the leader of the Western camp.

The economic establishments which have undertaken this large role in the oil world for about half a century are basically the international cartel companies, or "Seven Sisters". These are: Exxon (Esso), Gulf, Texaco, Mobil, (Chevron), Shell and BP (British Petroleum).

According to the size of capital, these seven companies are among the twelve largest industrial companies in the Western world. The following graded list[7] of these companies includes the seven cartel companies (bracketed):

| *Order* | *Company* | *Unit $1,000* |
|---|---|---|
| (1) | Exxon | 21,558,257 |
| (2) | Royal Dutch (Shell) | 20,066,802 |
| 3 | General Motors | 18,273,382 |
| (4) | Texaco | 12,032,174 |
| 5 | Ford | 11,634,000 |
| 6 | IBM | 10,792,402 |
| (7) | Gulf | 9,324,000 |
| (8) | Mobil | 9,216,713 |
| 9 | Nippon Steel | 8,622,916 |
| 10 | ITT | 8,617,897 |
| (11) | BP | 8,161,413 |
| (12) | Socal | 8,084,193 |

These seven companies were in 1972 producing 39.1% of oil inside the US, 77.6% of oil in the Middle East and Libya, and 77.1% of the total oil produced in the OPEC countries. Generally speaking, these companies, in addition to the French Oil Company (CFP) which is considered an Eighth Sister, were, by the end of that year, in control of 70% of world reserves (outside the European socialist countries and China). Their production comprised 65% of the total world oil production, their refining activities formed 55% of the entire world refining capacity, and they produced 60% of oil products in the world market.

According to the extent of their oil production, the "Sisters" may be put in the following order according to the total shares of crude oil produced in the world:

| | | |
|---|---|---|
| Exxon | 6,145 | thousand barrels/day. |
| Shell | 5,146 | " " |
| BP | 4,659 | " " |
| Texaco | 4,021 | " " |
| Gulf | 3,404 | " " |
| Socal | 3,323 | " " |
| Mobil | 2,399 | " " |

We have mentioned that the CFP may be considered an Eighth Sister, considering her giant size. She was added to the Iranian consortium in 1954, and her struggle for Algerian oil was behind numerous policies and international conspiracies. She has never stopped enlarging her activities in production and marketing in Europe and Africa. She may be described as one of the Sisters, although at the same time she provided a challenge from outside their circle.

The previous figures alone show the giant size of these companies, and the extent of their control over oil production and marketing. Through their complexity and the extent of their work and resources, these companies have become a part of an "international government". Their directors fly between Pittsburg and Iran, or between San Francisco and Oman with the same ease of someone moving across his own state. The computers working for these companies can analyse supply and demand in half the countries of the world. The administration can earmark hundreds of millions of dollars for the development of a new field, the construction of a port or the opening of a navigation line. Each of the Sisters is more than fifty years old, exceeding the age of most of the states they deal with. For several decades, the Sisters seemed like a mysterious and fearful world, to both producing and consuming countries. Their experience and connections, which overstep boundaries and pass over nations, were far stronger than the powers of any national government. Their income was greater than the income of most of the countries where they worked, and their fleets of tankers exceed in capacity any maritime fleet. They have owned and run entire cities in the desert.

In dealing with oil they were actually self-sufficient, immune to laws of supply and demand, and to changes in money markets. They controlled all forms of transaction within the circle of their business, and sold oil from one of their branches to another. Shell oil, for instance, was produced from Shell fields, to Shell tankers, to Shell refineries, then to Shell tankers which specialized in shipping refined products, and, through the Shell pipelines to Shell distribution stations. The Seven Sisters were in fact the first giant international companies in the world. According to the 1973 annual report of Exxon, the company was a multinational company at least fifty years before the term became familiar.[8] The Sisters extended their activities to other industrial fields covering petrochemicals, coal-mines and atomic energy.[9] They were not limited to energy production; they went into plastics in various forms and into fertilizers and drugs.

But oil in particular, with its wide geographical base and its political ramifications, was the source of the special influence of these companies, which made them master of half of the international trade. More important than that, oil gave these companies concessions through which some countries relinquished part of their sovereignty.

In spite of the competition, sometimes acute, among the Sisters themselves, they are still interrelated by joint projects and concessions around the globe from Alaska to Kuwait. The leadership of this group, throughout the last sixty years, has been in the hands of the two giant companies: Exxon and Shell.

Now we shall cast more light on the complexity of relations among the cartel companies, and on the weight of American interests within that cartel.

In the Middle East, and since the formation of the Iranian consortium (which included the Seven or Eight Sisters), the Sisters became interrelated in mutual projects which facilitated control and co-ordination in their work, as may be seen in the following sketch:

"The Sisters' Dance"

Mobil B P Texaco

| Socal | Bahrain P. Co. | Aramco | IPC | Abu Dhabi P. Co. | Iranian Cons. | KOC | Abu Dhabi | CFP |
|---|---|---|---|---|---|---|---|---|

Exxon Shell Gulf

The Sisters' Dance

The sketch shows the connections between the Sisters in the ownership of Middle East oil companies.

Source: Listening Sessions to statements by multi-national companies Vol.5 *(vide* Sampson, p.169)

In addition to the intricate and mutual interests which secure the control of joint work, all facilities and systems were designed to achieve this aim. The joint interests of oil men are connected with a network of communications, banks and accountants (specializing in oil accountancy and taxation).

Morgan Guarantee Bank offers securities to most of the Sisters (Exxon, Mobil, Shell and Texaco), while Chase-Manhattan and First National City bank deal with most of the external banking services, and are the centre of interest in the oil circles. The President of Chase-Manhattan, David Rockefeller, the grandson of the senior Rockfeller, is now occupied with financial affairs, and he is constantly the centre of attraction in oil circles. Though he is no longer a member of any board of directors in any of the Standard oil companies, his family still holds shares in all the Standard Sisters. David Rockefeller himself has close relations with the Middle East, which explains his frequent visits to the area.

Since tax evasion is the basic factor in the financial strength of oil companies, the role of accountants has become of pàrticular importance. Price Waterhouse has become a favourite with the Sisters. In 1971 this comany ran the accounts of Exxon, Socal, Gulf and Shell at the same time. The systems followed by this company are so complicated as to make one dizzy. When in 1973 Price Waterhouse published a sample of accounts systems followed in thirty oil companies, the variation was (in the words of Mr. Lisasian, Democratic Congressman for Wisconsin) such that no serious expert in business economics could reasonably compare the financial situations in any two of the thirty companies. It is known that tax evasion was the most important joint achievement in which the companies excelled. After this we may well imagine how the oil-producing countries can deal with these companies and calculate their dues.

One of the studies made of the Sisters came to the conclusion that the rates of oil supplies must be governed by some kind of tight planning, but the study adds that "there is no evidence whatsoever that this control is exercised by any system of central co-ordination"[11]. If such evidence is expected to take the form of an official declaration of an international cartel, then there is no such declaration. Yet central planning is not only indicated by rates of oil supplies controlled in international market; the official meetings of the companies in which the Sisters particpate form a clear indication of such planning. If we add to this the unofficial meetings, and all the other means mentioned above, the fact that planning does exist becomes more than obvious.

Concerning the relations of the American companies with the international cartel and oil production outside the US and the socialist countries, the following figures show how the percentages have developed:[12]

| Year | 1939 | 1950 | 1970 |
|---|---|---|---|
| American companies | 35% | 48% | 55% |
| English and Dutch companies | 49% | 43% | 26% |
| Others | 16% | 9% | 19% |
| | 100% | 100% | 100% |

The US is the largest country that refines and consumes oil and was, until 1974, the largest country involved in oil production too. Therefore, we find some companies limit their activities to the US. They are called independent companies and are in fact large establishments, approaching, in the size of their business, the status of the monopolies which are members in the international cartel. It is true that the Soviet Union is the biggest oil producer inside national boundaries, but when the size of oil supply which the American companies control is measured, whether from local production or overseas sources, the US remains the biggest producer.

Concerning oil production outside the US, the last table shows that the American companies were in control of 55% in 1970; and in the Middle East in particular the proportion exceeds 60%. The Anglo-Dutch Shell Company and the Anglo-Iranian Company (which became BP) represented the more serious competitors to the five American Sisters inside the cartel. But Shell also carries, within itself, American interests of considerable importance in the form of "US Shell Oil & Company". This company is so large that it forms more than one-third of the total income and gains reaped by the Shell group from activities all over the world. Moreover, the company experience in all branches of industry is used by international activities of the Shell group, so much so that the President of US Shell was made Administrative Director of the Shell Group. As for the non-American employees, their service with the American company at various periods is considered an important recommendation for promotion. According to all this, in technical fields as well as in organization and supplies, the industry has come under American direction.

These giant companies enjoy very high rates of profit. American oil cmpanies' investments abroad exceeded 5,000 million dollars (about one-third of American foreign investments). Profits remitted to the US in 1972 were estimated at 2,000 million dollars. This figure exceeds by 500 million dollars the size of the total cost of US oil imports and the amount of money spent by the US for the development of this industry abroad.

Concerning the Middle East in particular, the US Department of Commerce estimated, in 1970, the original capital in the oil industry as 1.5 billion dollars, yielding 1.2 billion dollars of profit, i.e. the investment return was 79% compared to 13.5% in production and mineral industries in the developing countries.[14]

The Fuel and Energy Director at the US State Department was quite right when he declared that the "loss" of an important part of these stocks becomes a serious question to the nation, as much as it is an invaluable loss for the private owner.

This declaration brings us in fact to the relation between the American state and the companies. The preceding figures show the importance of profits to the remitting companies as regards the American balance of payments. This matter exceeds the interests of the companies alone; it concerns the entire American economy, and ultimately is of strategic importance to the US government. But we must realize that the importance of the oil companies to the US government is not primarily assessed according to these narrow economic and commerical considerations.

The more serious and important question is the kind of product these companies are dealing with. The American government has been concerned with securing foreign oil resources since it became obligatory for the US to import increasing amounts of oil for local use. The US is also concerned with controlling oil resources to supply her allies in western Europe and Japan, and also to supply the Third World countries. Controlling these resources is a basic means to secure an international leadership of imperialist powers, and all this is realized through the Sisters and the Cartel.

> "The international oil industry is more than a mutual relation between supply sources and markets, or between companies and governments, or between companies and other companies. The system [of this industry] works through the framework of the larger and more complicated system of economic and political relations in the world."[16]

The monopolizing oil companies form a strong pressure group inside the American political establishment. Inside the group there are differences which, at certain stages, develop into acute struggles between the cartel members and the local companies working inside the US. Information about amounts of money spent by the oil companies to gain support inside Congress and the US administration is no longer a secret. The companies' participation in financing Nixon's election campaign in 1972 was fully published. The disclosure of these illegal offers caused a great shock to the American public, and recalled all the ugly memories about sabotage and corruption exercised by "oil money" throughout its history. In recent years there were recurrent disclosures about bribes made by the companies to responsible American officials and others abroad. The surprising and annoying thing was not the great amounts of "pay-off money", but the fact that these figures were so cleverly hidden within the companies' accounts. In spite of expert accountants, thousands of shareholders and various surveillance methods, these companies were able to set apart enormous amounts of money in their secret book-keeping labyrinths. Again, this indicates the difficulty in dealing with these companies and the impossibility of understanding the truth about their accounting systems.

These facts about the ability of the companies to exert power over political establishments, thus gaining a degree of independence to define their own course of action, do not rule out the fact that the companies are connected, and ultimately subject to the high strategic interests of the West and to the American state. Yet they are not mere administrative tools in the hands of that strategy; the companies form an influential dynamic power, endeavouring to increase their gains within the international economic system of the West, and within the limits of the strategy of the American state.

Various books and references abound with stories about the inter-relationship between companies' infiltration into the Middle East and the Arab area and between the policies of the Big Powers in the first half of the century. Each of these powers basically helped in opening the way before the companies, became connected with them, and supported the position of these companies using every means, from diplomatic connections to military intervention. It was no coincidence that the relative influence of the companies in the arena of international relations was in direct ratio at every stage to the political influence of the state to which these companies belonged. According to this rule, the American oil companies had the upper hand after World War II.

Insufficient evidence about the period since the Fifties has been disclosed compared to the period that preceded World War II. Yet this period concerns us particularly, since it is in this period that the importance of, and the war for, oil has reached a peak. Also this period is directly connected with our discussion of the prevailing circumstances which were influential on the Iraqi decision-makers in 1972. These are naturally connected with the development of the relations between the companies and the governments, and in particular with the American government, whether in the role played by these companies to serve American strategy, or the role played by the American government to support and protect the companies or the methods which regulate the relations of their mutual co-operation.

But the absence of evidence cannot justify Leeman's claim that "The American companies generally used to run their affairs in the Middle East by themselves",[17] nor does it justify the claim of the companies, at the same time as being involved in imperialist policies in their support of Israel's existence and expansion in the Arab area, that they are merely economic companies, concerned only with production affairs and profit-making, and that they are companies operating according to free market principles, and that they have no relation with their governments or the policies of those governments. These claims are scarcely credible. On the other hand, the absence of sufficient hard evidence does not mean the absence of sufficient indicative evidence.

Towards the end of Truman's term of office as President, the American government was openly preoccupied with securing the right of the Sisters to operate as a joint front. There was a sharp difference between the Attorney-General (who opposed the idea of the cartel) and the State Department. In a famous report, Dean Acheson, speaking for the State Department and supported by the Interior and Defence Departments, demanded strong government support for the Sisters, saying that "the companies play a vital part in providing the free world with one of the most important basic commodities" and "the operations of the American oil companies are our foreign policy tools for all practical purposes". The report raised a great amount of discussion when it was published, but the American administration insisted on supporting the Sisters in the face of opponents and in all cases brought against them, and demanded the formation of a committee to look into the interrelated differences among trust opposition, security and foreign policy. When the new administration took over after Eisenhower was elected, Secretary of State John Foster Dulles announced that all measures concerning the Sisters and oil affairs "must be directed in the light of necessary concern with all questions affecting national security".

The situation was fully decided from that date, when the Iranian question was at its peak. After this time the relationship between the Sisters and the government administration was organic.

And that relationship was not one of mere coordination between two independent systems; the administration inherited by Eisenhower, during whose office the coup against Musaddaq was effected, had put the State Department in the hands of John Foster Dulles, and the CIA in the hands of Allen Dulles. The first came from Sullivan & Cromwell Lawyers' Association, which represented Rockefeller's oil interests, the second had connections with the Middle East which predated World War I. Kermet Roosevelt, the CIA man who was directly responsible for the coup against Musaddaq, left public service after the success of his operation, and became Vice-President of (Sister) Gulf Oil. The US State Department sent Herbert Hoover Jr, after the coup, to formulate new relations with the Shah's government, after which the American companies entered into the Iranian Oil Company, in 1954. In addition to representing the State Department, Hoover used to work for one of the oil companies as a consultant. Senator Kieffauvcr spoke about this period (three years later) saying that "all Iranian affairs and the path of government policy were dictated [by the companies] to a large extent, including the policy of Mr Hoover, who was a spokesman for the government".[18]

After the nationalization of the Suez Canal (1956) the American oil companies and government were highly upset. One of the Standard Oil of New Jersey periodicals (Exxon) wrote at the time:

> "Every now and then, a Middle East politician appears with the hope of securing or holding office on the basis of hatred of the foreigners. Such men do not hesitate to abolish the holy agreements concluded with the citizens or governments of other countries. It is useful, for instance, if the US could issue a statement (alongside other statements about supporting independence and liberation from foreign rule) of a general and firm nature, running roughly as follows: Any government, unilaterally terminating a concluded agreement, may expect from the US alone or in collaboration with other countries or through the UN the undertaking of economic measures, and the abolishment of all agreements concluded with the aggressive nation in the manner which the US finds the circumstances call for. Such a statement would show other nations what we are ready for, and would let them realize that international agreements or private contracts cannot be handled or abolished lightly or hastily merely if such a step seems useful to those nations".

The eldest Sister here clearly expressed the relationship between companies and governments, and the fact that they could enforce their exploitation terms under the direct protection of these companies, which in turn serve the government policies. On this point the agreement was complete between the Sisters and the American Government. On 12 August 1956, a secret meeting was held between the representatives of both sides where Dulles announced: "... Therefore, any nationalization of this sort of possessions of international importance, is an issue much greater than any compensation for share-holders, an issue which calls for international intervention".

After the defeat of the triple aggression on Suez, Eisenhower presented his doctrine to "fight international communism" and to "fill the Middle East vacuum"; the doctrine smacked of oil. It was another step taken by the American state to protect oil companies. Senator Carroll (of Colorado) stated "evidence points to the fact that men in office and lawyers in these huge corporations [meaning oil companies] have sat with government administration to help draw up the policy which we have come to discuss in this assembly". Senator Kieffauver added:

> "I think it is unfortunate that information about those meetings should remain a secret, kept away from the people. However, I think it can be said that in the case of all these important government meetings, even those attended by Mr Dulles, or those several meetings, attended by Mr Herbert Hoover with leaders of government interests concerned, and with the representatives of huge international oil companies... in all these meetings held to discuss this programme [meaning the Eisenhower Doctrine] there were no minutes taken of the proceedings, and we were unable to get any such minutes. Even though it has been customary to provide minutes and outlines of any important meeting, and even of small and unofficial meetings, they have very carefully avoided the preparation of any minutes or outlines telling of what had happened in those meetings, and we had to follow a difficult path to reach such information, and pick up separate data here and there in order to form a correct picture".

And what was the picture obtained by Kieffauver at that time?

> "There are very numerous indications that the material of this programme was cut out primarily to fit the needs of international oil companies. And, instead of the oil being one factor in the formulation of foreign policy, it became the influential power controlling such policy... During the last weeks a question emerged about the real necessity and aim of this resolution about the Middle East. There are really two aims hidden under the superficial aims, which the resolution serves: the first is to warn the Arab countries against nationalizing the oil concessions offered to American oil companies; the second is to allow for immediate intervention when such nationalization takes place without any delay imposed by opposition, or by discussion in Congress".[19]

In 1958, when a patriotic force led a revolution in Baghdad, Professor Leeman wrote:

"The question raised at the time was: what is the policy which the West should follow in the event that the revolution of Iraq should extend to Kuwait, which is under British protection, or to Saudi Arabia? Or, if efforts were made to merge those two states in a larger Arab formation, which may become a united Arab republic under the presidency of Gamal Abdel Nasser? The British have required American assurance to join them in using force if necessary, to keep the Arab Gulf oil-producing states separate and independent."

Leeman also says that results of diplomatic communications between London and Washington in that period (July-August 1958) emphasized the necessity of preserving relations existing with Kuwait and other Gulf States.[20] Incidentally, Leeman should have known what he was talking about, for he was well known for his close relations with cartel companies.

During that period, the reader undoubtedly remembers the movements of the Sixth Fleet and the landing in Lebanon. Lebanon was an established centre of oil and economic interests, and the main centre of CIA activities. Miles Copeland confirms that oil companies had become – before the Lebanese crisis – quite influential in Middle East policy, heading towards the use of violence. The major oil companies opened special offices in Beirut. (Copeland himself – the CIA man – resigned his official job, and set up in May 1957 a consultation office for one of the oil companies. These offices had to watch the exploding situation closely, and these efforts used to help "the Government Liaison Office" attached to the Tapline Company, which was described by Copeland as highly efficient.) The question was raised about US marines landing in Lebanese territory, in compliance with the Eisenhower Doctrine, but the US government hesitated in taking a decision. But after the 14 July revolution, oil circles were "shocked by terrible reports coming to Beirut from Baghdad," and so it was decided to support Chamoun's claim about the necessity of American violence against Lebanese patriotic forces, to serve as a warning to the revolution in Iraq. When the Sixth Fleet commander arrived in the US embassy in Beirut, he announced: "We have stopped an economic recession in the US".[21]

When J.F. Kennedy became President of the US and readjusted major appointments, he chose John McCowen to become CIA Managing Director. In discussing this appointment in the Armed Forces Committee in the US Senate, McCowen said (January 1962) that he had been Director of the "Panama Pacific Oil Tankers Company." and that he had bought Standard Oil of California Company bonds for one million dollars. (It is known that this company operated in the Middle East, Indonesia and Latin America.) After McCowen had presented these "recommendations" for appointment in the new post, Senator Joseph Clarke commented by saying that "every American, who is familiar at all with the facts, knows that American oil companies are deeply involved in Middle East politics... and that the CIA administration is also deeply involved in those politics."[22]

Yet, it was well known that Kennedy was not in complete agreement with the oil companies. When he was a Senator for Massachusetts, he played a prominent part in the bloc which included the so-called energy consumers, who opposed the producers. The struggle on both sides was basically about limitations on oil imports. When Kennedy was assassinated in Texas (22 February 1963) some accusing fingers were pointed at the Sisters. Whatever truth these accusations may have, the certain thing is that the oil industry rejoiced at the news. As soon as Johnson assumed office, he stopped all the tendencies initiated by his predecessor. Incidentally, Robert Kennedy, who in 1964 became Senator for New York, was active in the same direction as his brother, J.F. Kennedy, that is, against any limitations of oil imports which would raise energy prices for consumers in the US. In a commentary on major nominees for the US Presidency, *Oil and Gas Magazine* said that Kennedy might be a danger to the oil industry, while Nixon might be a friend, if his previous statements were taken into account. So, Robert Kennedy's assassination in California (June 1968), after winning the primary of that state, was received with the same joy as his brother J.F. Kennedy's assassination.

Incidentally, Edward Kennedy made a speech in December 1968, sharply criticizing the US oil industry and accusing it of surpassing the bounds of selfishness, and demanded that it "submit the entire oil import limitation programme to the US Supreme Court to decide its legality". The young Senator continued his struggle with the oil monopolies when he began to be considered as a possible nomination for the Presidency in 1972. Then he was involved in that car incident in July 1969 in which one person was killed, which practically cancelled the possibilities of his nomination. It is also interesting to note here the participation of the giant oil companies in financing Nixon's election campaign.[23]

Those who called for facilitation of oil imports argued generally about the economic interests of US industry, since imported oil costs less. The oil producers argued that easing limitations on imported oil would put the US oil industry in a weak competitive position, leading to greater US dependence on growing quantities of imported oil, which would harm the self-sufficiency objective and threaten national security. We are not concerned with a detailed discussion of this controversy; what is important for us here is to realize that the oil issue is always connected with the highest levels of US policy, and that differences of attitudes – even inside official establishments – may be solved with bullets.

This battle of the Sixties inside the US was connected with another battle fought by the cartel companies with the US government's support and protection in western Europe. Since the late Fifties, the cartel companies had begun to realize the danger of competition from national European companies, and from Soviet oil sales in Western markets, eroding their traditional influence. The great shock came in 1960, when the Soviet government concluded a deal with the Italian government for 100 thousand barrels per day in return for Italian artificial rubber and pipeline equipment and 40-inch diameter pipes, which were to be used to transport Soviet oil to western Europe. The deal met 20% of Italian oil needs and was economically profitable. But Mr Rathbone (Chairman of the Exxon Board of Directors) said that the deal was a political act designed to create trouble. The *New York Times* said the deal threatened the security of the Western world. But Italy was at the same time concerned with the development of her independent production. It is said that the US Secretary of State, J.F. Dulles, had a heated argument with Mr Gronchi, the Italian President, and reminded him that US aid was not offered to Italy to enable its government establishments to compete with Standard Oil and other cartel members. The US government took a strong stand and refused financial aid to Italy which would support the national oil industry and help its development. The US government raised the question inside NATO too, but the Italian government continued its policy.

The Italian oil establishment (ENI) was headed by a dynamic patriotic personality: Mattei. Under his leadership, ENI was able in 1961 to extend its operations from Italy, Egypt and Iran to Somalia, the Maghreb, Tunisia, Greece, southern Germany, Guinea, and then Libya. Mattei said:

> "The power of the large international companies, exercised in the refining operations and the distribution of their products in Italy,

enables those companies in fact to deprive consumers and the entire Italian economy of the full benefits of competition, which is increasing every day among oil producers, and which is influential in deciding basic reductions in price. That is a fact which is now clearly realized by other countries too. Those countries have come to believe that a state establishment is the best weapon with which to stand up to the powerful influence of international groups. Nowadays, real possibilities are available to us to set limitations to and set up obstacles against greater exploitation of energy resources in the country; and this can be done even on a larger scale, overstepping national boundaries, by, for instance, the European Common Market, where Italy is not alone in having such aspirations."

The end of Mattei's story was not surprising. The plane on which he was travelling suddenly exploded in October 1962.[24]

Coming back to the Middle East, it is certain that co-operation between the US government and the Sisters was at its peak in planning the June War of 1967, but the details have not seeped out yet. On the contrary, we are told by all the media that the companies were quite free of political involvement with the US government. It is no secret that the US President at the time of the aggression, Lyndon Johnson, had been, for twenty years, defending the interests of oil men in Texas (when he was a Senator), and that Foreign Secretary Dean Rusk had worked for the Rockefellers. It is also well-known that the future of the oil monopolies is dependent on the blocking of the Arab liberation movement. Senator Jacob Javits of New York (who has close connections with Nelson Rockefeller, New York Governor and prominent oil man) met with Dean Rusk, a little before the 1967 aggression. He then published a statement threatening Egypt with military measures, saying that the US could, in collaboration with Israel, undertake military action against the Arab countries. In July 1967, Javits hurried round to his friends at the Capitol and after securing a large number of votes, he submitted to the Senate a draft resolution supporting US policy in the Near and Middle East.

The fact that the special relationship between the US and Israel is basically affected by the oil question needs no lengthy explanation. And the fact that the Sixth Fleet and the surrounding bases provide protection for oil companies is again a question which needs no lengthy explanation. When Nixon visited the Sixth Fleet in September 1970, the *New York Times* said that the visit was an expression of the US government's intention to "allude to the use of force in order to protect her oil interests".

In spite of this, the Sisters doubled their publicity efforts after 1967, and unceasingly repeated that they were independent companies, whose concerns did not exceed the economic interests of their share-holders, the producing states and the consumers. But this absurd claim found some credence among responsible Arab officials some of whom insisted that oil was independent from politics, and that because the companies did not subject oil to US policy, then the Arab states also should not subject oil to their national policies of liberation.

In any case, the companies during this last period were insisting on their independence from their states, while it was known that they frankly relied on Western governments, and the US government in particular, in every crisis with a producing state, whether it be Iraq, Iran, Libya, Algeria or Venezuela. They employed every means of pressure that the US government had at their disposal against OPEC, in order to enforce a version of co-operation between the Sisters and the major and conservative oil states inside that organization.

At the same time, and under growing pressure from OPEC states, planes carried 9,000 marines with their heavy arms, from their bases in North Carolina to a place in the desert near California, for "war games" in a summer heat reaching 110°F. There, they divided themselves into two teams: the US team (leather collars) and the Yarmonians (armed with Soviet weapons). And they began the game.

The 'Yarmonians' – according to news reports – wore uniforms very close to those of the Libyan army. The US team charged first with the soldiers shouting: "Come on kids! Let's take that oil! " Colonel Leary, who was in charge of the training, explained the aim of these manoeuvres by saying that "the Pentagon has plans to invade all civilized countries of the world. The Middle East, as is obvious, is like a barrel of gunpowder; and we would be fools if we were not on the alert".

The US press gave full coverage to these manoeuvres, in order to get the message of the Sisters' importance in US strategy across to the oil states.

Again, the news released was undoubtedly incomplete. A week before the publication of that news, the American aircraft carrier *Constellation* was sailing in the Arab Gulf for the first time in twenty-six years. The *Christian Science Monitor* commented on the incident by saying that "The visit proves that the US cannot accept any threat to the stopping of oil imports from the Arab Gulf states".

In the second week of December 1974, a force of 2,000 marines of the Sixth Fleet landed on the shores of Sardinia in a monoeuvre modelled on the invasion of Arab lands. Admiral Turner, who was in charge of the manoeuvre, commented that: "We do not want to invade the Middle East, but we have to be ready for that".

On 13 January 1975, Kissinger himself stated to *Business Week* magazine that US military action might be necessary if Arab countries should threaten the industrial world with real suffocation.[25]

These facts belong to a period following the Iraqi nationalization of oil, but their significance was not lost on those who were thinking of nationalization in 1972. Even after all this publication of government support, the Sisters maintained their insistence that they were independent. The situation got to such a state that Aramco was even accused of faithlessness to American interests; that is to say, the company had lost even her independence, and had become a tool of Saudi Arabia, when the latter's interests conflicted with US strategy.

**Two: Tragedy of the First Confrontation**

The previous facts aim to show that oil is an explosive issue, and that an oil-producing state aiming to control her oil takes the risk of running up against fearful international stratagems and organizations larger than the state itself. Behind these establishments – organized in the cartel – there is also the power of the US itself. Iran was the first state to attempt a confrontation. The attempt reverberated around the world.

The tragic failure of the Iranian nationalization experiment was an example of the "perfect crime" committed by the imperialist powers with exceeding brutality. It became a nightmare, terrifying when mentioned, and a warning to anyone who thought of repeating the experiment.[1]

On 28 April 1951, the Shah of Iran appointed Dr Muhammad Musaddaq Prime Minister on the recommendation of the House of Representatives. Musaddaq did not accept the appointment until he had secured the approval of both the Representatives and the Senators for the policy of the immediate take-over of the Anglo-Iranian Company, in accordance with the Nationalization Law; and the two Houses gave him unanimous approval. On 1 May the Emperor signed the Nationalization Project which became law, thus ending fifty years of

British control. Lord Strathalmond (Chairman of the Board of Directors, Anglo-Iranian Oil Company), cried: "It is a blow. It is really terrible to wake up one morning and find that you have lost most of your oil possessions". Morrison (the British Foreign Minister at the time) stated: "We cannot accept a radical change in the company situation in Iran by a unilateral procedure".

The war had begun.

At first, legal methods were employed. The company sent a memorandum asking for referral of the issue to arbitration, in accordance with provisions of the Concession Agreement of 1933. The British ambassador, in his turn, presented a British government memorandum, whereby the government reserved her right to refer disputes to the International Court of Justice in The Hague, if the Iranian Government did not accept arbitration.

The British government memorandum was in fact a declaration that counter attack command was being directed from Downing Street, whence instructions were issued to the Anglo-Iranian management to stop payment of dues to the Iranian government. The Iranian government answered by rejecting arbitration, on the principle that nationalization of the Iranian oil industry was not an issue subject to arbitration, and that there was no international court with a power to look into this issue, which was the immediate concern of the Iranian government. Musaddaq stated on 28 May that "When the government behaves within her rights of sovereignty, which may entail some harm to a private company, the only solution which that company may resort to is to ask the government for compensation. That compensation is provided for in the Law of nationalization. The 1933 Agreement is not under discussion, and any mention of that Agreement is out of place". Throughout the country there were demonstrations in support of the government.

When the British sent a division of paratroopers to Cyprus, to be ready to protect British lives and possessions in Iran on 25 May and the British Army units in Iraq were put on the alert; orders were issued to the Iranian Army to be ready for the defence and if necessary the destruction of oil installations.

On the other side of the Atlantic, the US was preparing her own game. She did not approve of the British military intervention and on top of this Soviet troops were gathering along the northern borders of Iran.

The British government finally reviewed her position and gave up her attempt at armed intervention.

On 14 June, a first round of talks was started between a delegation from the nationalized company and the Iranian government. The company's strategy was based on the assumption that Iran was in a weak position, because she needed to sell her oil; and the Iranian government imagined that the company was in a weak position because the West could not do without Iranian oil. Events proved that both sides had miscalculated, because Iranian oil though important could be replaced by the surplus from neighbouring Middle Eastern countries. Nor did Iran collapse when the export of oil was stopped. She persevered for three years, despite problems, because the Iranian economy was not greatly dependent on oil.

In any case, negotiations failed on 19 June because the company's proposals, as previously explained by Musaddaq, ran counter to the Law of Nationalization, and instructions were issued to start the takeover process of company establishments although the full implementation of these instructions was postponed. On the Company's side, instructions were issued to tankers to refuse oil against receipts recognizing Iran's possession of her oil. And on 26 June, Anglo-Iranian representatives ordered the tankers to unload and leave Abadan at once. On the same day, the battleship *Moratius* was heading to the Gulf to evacuate British subjects. The 3,000 British technicians refused an offer from the Iranian authorities to continue work on new and individual contracts with the nationalized company. The fields began to stop pumping, one after the other, and with the cessation of export, work stopped completely at the Abadan Refinery on 31 July; thousands of workmen and hundreds of Iranian engineers were laid off.

During this time, the US began open intervention. Averell Harriman headed a special delegation sent by President Truman on 15 July, and Musaddaq's reception of the US delegate was the first occasion for disagreement inside the front supporting nationalization. The security forces collided with demonstrators and national newspapers were confiscated. But the Harriman talks led to a second round of negotiations with the Anglo-Iranian Company, a second round, which failed for the same reasons as the first – the proposals ran counter to the Law of Nationalization. Harriman wrote to Musaddaq on 21 August saying he believed the British proposals were in accord with the Iranian point of view.

Faced with the obstinacy of the Iranian government, Anglo-Iranian strengthened her embargo on Iranian oil with the help of her Sisters in the cartel – issuing a warning to every company or individual who came to buy nationalized oil, announcing that the company would take every measure to secure her rights. The US government helped by threatening every importing country with withdrawal of US aid. In spite of all this, the Iranian authorities made great efforts to resume production after fully taking over the industry in October 1951, and it actually became possible to resume work and partly operate Abadan Refinery. But the problem was export. The Italian ENI used to import the majority of her oil needs through the Anglo-Iranian, and Iran tried to conclude a direct agreement with that establishment, but failed (through pressures exercised by the cartel). This failure to export continued for a long period. From the wells of other oil-countries, the cartel companies helped to replace Iranian oil, in spite of growing international demand. So the oil production of 535 million tons in these countries in 1950 increased to 637 million tons by the end of 1953. In the same period Iranian exports decreased from 54 million tons to 132 thousand tons in 1952 and 1953.

Britain demanded that the dispute be submitted to the Security Council in 1951. Musaddaq headed a large delegation to New York and among his papers were documents which his government had seized at the house of the Anglo-Iranian President in Tehran. The documents disclosed the company's interference in Iranian politics, by appointment and dismissal of ministers and consecutive governments, and by spending huge sums of money to support the company nominees in the elections.

These documents were important to all oil-producing countries as much as they were to Iran. In fact, Musaddaq achieved a great international success when Britain failed to secure a UN resolution in her support. Before that, Musaddaq had gone to the International Court of Justice, and had scored an even greater success for the principle of nationalization, as the Court issued a decision in July 1952 stating that the dispute was local, and fell within the jurisdiction of the Iranian state.

The embargo continued... and conspiracies seemed likely to break the solidarity of the government, using existing differences, which Musaddaq did not handle with efficiency and decisiveness, in spite of his great popularity and power. He had all the patriotic forces behind him, with the Shah as the head of a counter group. Musaddaq took powers from the House to purge the government system. He started with the sensitive departments, conducting new elections in which his party "The Patriotic Front" won the majority of seats. He wanted to keep the Ministry of Defence in the new cabinet. But when the Shah refused, Musaddaq submitted his resignation in June 1952 and the Shah appointed Qawam Al-Saltana to succeed him. The country was overrun by a massive revolution which almost uprooted the Shah himself, but it was Musaddaq who stood by him in order to fulfil his oath to be faithful to the throne! The revolution only achieved Musaddaq's return as Prime Minister and Minister of Defence. And he began a limited campaign to demobilize several high-ranking officers.

But the conspiracies began to concentrate on dispersing the Patriotic Front, and they succeeded.

After the separation of religious leaders from Musaddaq, another ally to be wrested from him was General Zahidi, who was trusted by Musaddaq and had been given the Interior Ministry. Moreover, the reactionary forces effected further divisions within the Patriotic Front Party, which led to dissent among other prominent members. It was only then that Musaddaq realized the danger from movements opposing his government and his nationalization battle, so he requested the Shah to leave the country for a period of time.

It is reported that Musaddaq said to the Shah that he had no ambition to become a president of a republic, nor to remain in office, and that when the oil crisis was solved he would leave his position to another person. It is also said that he told the Shah that he was under oath to be faithful to him, and that Musaddaq could not forget the Shah's favor when he was the Crown Prince: Musaddaq had been in prison then, and the Shah had saved his life.[2] One tends to believe this story as it is in keeping with the mentality and personality of the 'knight' who led the nationalization campaign. It also is in keeping with the scenario of events which followed. The Shah consented to the Prime Minister's demand, and 4 February 1953 was the date set for his departure. But demonstrations were organized that day asking the Shah to remain. A letter arrived from Kashani – as head of the House of Representatives – asking the Shah to cancel his departure. This was supported by another letter from Bahbahani, emphasizing the same request. And so the departure was cancelled, and Musaddaq was almost killed that day. Communication was completely severed between the Prime Minister and the Emperor. Although this situation led to a complete division in the state leadership, and created a situation of complete chaos, Musaddaq, the ideal liberal, refused to listen to advice from his supporters when they told him that the country was in a state of war, and that such a situation could not be handled with democracy. "You have to hit with an iron hand at your enemies and the enemy of the people before the last chance is lost." But he answered, with a Socratic spirit, that if he violated the law he would be like them, and then there would be no difference between the faithful and the traitors.[3]

In reality, the national rule formulated by Musaddaq was – by its very formation and system – unable to continue the revolutionary attitude which was started by the nationalization decision. There is no doubt that Musaddaq, with all the historic significance and popular feeling that he represented, and with all his courage and sacrifice, was a centre of both strength and weakness at that moment in history. The man's strength was a myth, and he faced the most dangerous challenges and waged a fierce battle even though he was an elderly man of seventy. He was of fragile health, hardly able to walk; but when he spoke to the masses he pulled himself together and gave vent to all his emotions, never ashamed when his voice trembled and his tears flowed while speaking about the affairs of his country. This man was able to recruit the Iranian masses in reverent resistance. But since extensive battles cannot be conducted by myth alone, the point of weakness in his leadership was due to his political concepts, psychological formation and family and class connections. It was difficult for him to overstep all that when he was at that age. Dr Musaddaq was descended from an old family of great power and wealth. He began his political career when he revolted against his wife's maternal uncle, the despotic King Muhammad Ali. He joined the activities which led to the downfall of the King and his escape from the country. He lived all his life in struggle, always working for the democratic life he envisaged. All the violent shocks during the nationalization battle could not convince him that the situation needed a tool stronger than the brittle tool of liberalism.

The situation did not change when General Anshartous (Director General of Police) was assassinated in June 1953, only three months before the fall of the national system. It became clear that the conspiracy which led to the kidnapping of Anshartous (Musaddaq's close friend) and his butchering to pieces was a large conspiracy, including Dr. Baqa, an MP and old friend of the Prime Minister. It was natural that the cabinet should suggest trial by court martial, because the crime affected the sovereignty and security of the country. But the elderly leader obstinately refused the proposal, and referred the criminals to the Ministry of Justice, in whose corridors the crime was lost, and the investigation petered out.

Similarly in international relations, the Iranian leadership for a long time exaggerated the possibility of exploiting differences between American and British interests. True, the differences were there, but it was more a case of the Americans wanting a portion of the English cake, and not a case of either supporting Iranian interests, embodied in the idea of nationalization which would infect a strategic area with what the Americans feared would be an epidemic of contagious nationalization.

To briefly divert from the issue of oil nationalization – despite its great importance – we must not forget the severity of the Cold War in the early Fifties, and the sharp polarization between Eastern and Western camps. It is certain that the US was not ready to solve any differences with Britain in a manner disturbing to the Western alliance, so important in facing up to the Soviet Union. According to ideas prevailing at that period, Iran was supposed to ultimately conform to the role assigned to her by American strategy, and, under US control. There was no room for a "non-aligned national" attitude, Iran had to be either with the US, and on US terms, or crushed. True, Iran has common borders with the Soviet Union, which could be used as a bargaining card. But the possibility of using the Anglo-American differences, in the circumstances of the Fifties, remained of limited value.

On 29 June, Eisenhower sent his reply to Musaddaq's demand of a loan, with a clear warning that either the Iranian government immediately agree to start talks about the future of oil, or there would be no loan. The experienced old man was ready for the answer. He held a conference round his sick-bed and swore that he would never give in to the imperialists, who were trying to insult the Iranian people.

In the midst of all this economic turmoil, political laceration, and duplicity of authority, the CIA hastened to use its tools. The operation was assigned to Kermet Roosevelt, who had been director of the Agency operations in the Middle East. Roosevelt entered Iran in the normal manner, and rushed by car to Tehran, but soon disappeared. He had to do so as he was a recognizable face in Iran. He was obliged to move his headquarters more than once, as he did not enjoy US embassy protection, but was helped by about five Americans, some of whom were CIA administrators attached to the US embassy, assisted by seven local secret policemen, two of them senior Iranian security men. These two men communicated with Roosevelt by proxy, but never set eyes on him during the entire operation.

During this time Swartzkopf had to be in Tehran and get in touch with his friend General Zahidi and his other assistants. At the same time Allen Dulles became CIA Director, went to Switzerland and met with the US Ambassador in Tehran and with prominent Iranian personalities. Then events followed quickly: the Shah left the country (on 13 August) after the failure to remove Musaddaq as Prime Minister, and in the midst of resounding demonstrations of support, Dr Fatimi asked Musaddaq to initiate quickly a move to oust the Shah and appoint a crown prince. Dr Fatimi said: "Give us arms; every hour you delay in doing so will push the country a hundred years back, and will cause our total loss... Save yourself and save us before the catastrophe".[4]

Unfortunately Musaddaq did not respond... and on 17 August Kermet Roosevelt gave his orders, from his hideout; the conspiring forces advanced, and down came Musaddaq.

Musaddaq fell and with him fell his experiment... and Iranian society was deeply disturbed, and all its social and political structures were in turmoil for several decades. The enemies had no doubt possessed several trump cards, but the national administration had had its cards too: the moral energy of the Iranian masses was able to face any economic problem.

The Iranian experiment failed because the battle was highly complex. Unfortunately, the Iranian leadership could not use its cards with the same dexterity with which the enemy was playing. Definitely, the defeat was political and not technical or economic; and the reasons were internal rather than due to external intervention. Or rather, the external intervention could not have succeeded without the presence of the internal reasons.

An understanding of this outcome is highly important, and I have deliberately treated the question extensively to underline its significance. I hope that this Iranian picture, with all its internal contradictions and complications, will be present in our minds as we follow up what was later achieved in the experiment of the Iraqi nationalization. The Iraqi leadership closely studied the reasons for Iran's defeat. And, if the result of the Iraqi nationalization was different, the difference must have been basically caused by political and internal reasons.

In any case, and whatever the analysis, the oil companies of the Middle East considered the nationalization of the oil industry in Iran in 1951 as a turning point. *Desert Enterprise* magazine wrote: "Something wrong had happened, and it seemed as if the existence of Western companies in the Middle East had suddenly received a violent knock-out blow. It was now of vital importance for the companies to see that what had happened should not happen again anywhere else".[5]

It may be said that the Seventies are different, and that the distribution of power in the oil industry is no longer the same as in the early Fifties. Undoubtedly this is true to a large extent. The new developments may be outlined as follows:

1. The emergence of national companies in the fields of production and marketing, the growing importance of these companies, their role in oil relations and the effect on markets and consuming countries, in addition to their role in the producing countries, and the role they are preparing to play in the future.

2. The emergence and growing importance of companies dissenting from the monopolies, and the role they play in the international oil market, and in their participation in production in some of the producing countries, and ultimately the effect of all this on the traditional oil power structure.

3. The growing part of the Soviet Union on the international oil market, and in helping producing countries to set up oil industries separate from the monopolies.

4. Political changes on the international scene, and the emergence of nations with a more powerful part to play in international relations, which means a change in the oil power structure.

This survey of the development in the oil world which was initially presented by Dr Abdul-Rahman Muneef[6] leaves little for me to add, except the distinguished part played by OPEC, which I shall take up later on in this book. But first I should like to shed more light on the first and second points.

Concerning the first point: the growth in the part played by national oil companies called for the need to revise the concessions held by the big oil companies. The year 1970 witnessed the successful Libyan confrontation with the oil companies operating on Libyan territory. Other similar confrontations in other oil countries followed. In the same year, Algeria nationalized the shares of the non-French companies, and the Franco-Algerian negotiations began to revise the 1965 agreement. Libya nationalized the marketing facilities. In Iraq the national investment for the development of North Rumaila field began. In Venezuela the income tax on companies went up. In February and April 1971 Algeria nationalized 51% of the French Oil Company. In December, Libya nationalized BP Shares at the Sari field.

As for the second point, the growing importance of companies dissenting from the cartel, these companies represent a conflict between independent American companies and the seven cartel companies on the one hand, and on the other the competition between the combined power of all the American companies and the interests of large consuming countries.

The so-called independent American companies are in fact companies not included in the Seven Sisters family, i.e. companies with no international system of production, transport, refining and marketing. Most of these independent companies have reached a large size and attained huge wealth by the standards of any industry. Therefore, most of these independent companies are considered, inside the US, as giant companies equal to the Sisters. With the entrance of some of these Companies into the cartel with the Iranian consortium, a sharp competitive factor emerged, and gained momentum with the entrance of the consortium into Libya.

The expression of interests on the part of consuming countries began in many countries with the formation of national companies pumping oil, though only on a small scale. This happened in Japan, France, Italy, Finland, Norway, Austria, Spain and Portugal. The consuming countries also started to try to convince the cartel companies to follow a policy in refining and marketing which the states concerned consider nationally desirable.

Undoubtedly, the entrance of these companies into oil production helped to bring about significant changes in the system of relations between producing areas and the cartel companies, and, more exactly, helped the producing countries to gain better terms from the cartel companies. The consuming countries and their independent companies were ready to make large concessions to producing countries in order to secure for themselves a degree of liberation from the cartel companies. "We have had enough of being under the thumb of the companies and for many reasons, especially because these huge companies carry out their operations on an international scale, and because they are Anglo-Saxon. This feeling is becoming stronger with a number of European states to such a degree that the situation for these states has become related to their national security." These are the words of

*Petroleum Swiss Service* magazine, the official organ of businessmen in member countries of the European Common Market.

This struggle between the independents – American, European and Japanese – and the cartel resulted in a development in the relative value of each of the teams in oil production operations (outside the US and the socialist bloc) which is shown in these figures:[7]

| Year | 1950 | 1960 | 1965 | 1970 |
|---|---|---|---|---|
| Cartel | 88 | 78 | 75 | 70 |
| Independents | 12 | 22 | 25 | 30 |
| | 100% | 100% | 100% | 100% |

These developments are undoubtedly positive. And the Iraqi Command had undoubtedly taken them into consideration while taking the decision to nationalize IPC. But undoubtedly again, those who hesitated realized, in their turn, the implications of these developments, and knew that while they could decrease the difficulties which the Iraqi experiment would face, yet they could not abolish those difficulties. The danger they represented was still great, and this was later proved.

The developments enumerated do not rule out the fact that the US is still a power capable of force and conspiracy. And the cartel, despite the changes in its power was still in command of 70% of the oil circulating on the international market. The largest independent oil companies in western Europe and Japan are still of limited size when compared to the Sisters. While the Sisters are among the largest 12 industrial companies in the world, we find that the largest west European and Japanese oil companies come – in the size of stocks – after No. 20 in the list of giant industrial enterprises. If we compare the product of the largest eight west European and Japanese companies we find out that the total product of the first group is thirty times more than the product of the second group.

The enemy then was still in command of the tools which he used in Iran and succeeded with. And also to be considered was the fact that the struggle with the independent companies or the French company was less hazardous than struggle with IPC. This company was the first consortium of the oil companies, and it represented in its structure five of the Eight Sisters.

Added to this was the importance of Iraq in the oil world – in the geo-political and political sense – and the fact that Iraq had also planned a full and not a partial nationalization. This would undoubtedly double the enemy's intention to attack.

I hope that I have succeeded in this chapter to answer the first question posed at the end of Chapter I: Why did all parties fail to forsee the Iraqi nationalization? I hope I have also explained why nationalization was a difficult decision.

Now we move to the second question: at the end of Chapter I: was the nationalization decision a leap in the dark? Or was it an uncalculated risk which met with good luck?

In fact, the answer to this question forms the remaining chapters of the book.

**NOTES**

Part One

1. *Vide* M. Ajlan, *Oil and the Arabs* (Beirut, 1974) pp.19-27 (Arabic).
2. P.R. Odell, *Oil and World Power, Background to the Oil Crisis.* (Penguin Books, 1975) p.9.
3. C.P. Dalton and T. White, *Petroleum Review*, June, 1973.
4. Q.A. Abbas, "Oil and Energy in Socialist Europe", *Oil and Development Review* I,4 (1976) (Arabic).
5. International Arab Labour Union, *Arab Oil Workers and the Oil Battle:* Seminar on Oil, Baghdad, 1972 (Arabic).
6. P. Dumas, his fifth address to the Americans, in P. Fontaine, *New Race towards Petrol*, Arabic translation by F. Abdul-Majeed and A.B.M. Bakr (*Our Choice* Series no. 171) (Cairo, n.d.) p. 11
7. *Fortune* magazine, May & September 1973.
8. A. Sampson, *The Seven Sisters: The Great Companies and the World They Made* (London, 1975) pp.5, 13, 32, 42, 188 and 201.
9. J. Stork, *Energy Crisis in the US and Middle East Crisis* (Beirut, 1974) pp.95 and 101 and table on p. 98 (Arabic translation).
10. Sampson, *op. cit.* pp. 204 and 205.
11. E. Penrose, *International Petroleum Industry* (London, 1968), p. 152
12. *Oil & Petroleum Year Book*, 1940, 1971.
13. Odell, *op. cit.*, p.21.
14. A-A. Kubba, *OPEC, Past & Present* (Vienna, 1974) p.78.
15. Odell, *op. cit.*, p.23
16. *ibid.*, p.166
17. W.A. Leeman, *The Price of Middle East Oil*, Cornell University Press, 1962, ch.7
18. A. Hussein, *op. cit.*, p.62.
19. *ibid.*, pp.68 and 73.
20. Leeman, *op. cit.*, ch.10.
21. M. Copeland, *The Game of Nations* (London, 1970) ch.10.
22. D. Wise & T. Ross, *The Secret Government*, (Cairo, n.d.) Arabic translation by G. Azeez, pp.247-8.

23. B. Ratchkov, *Oil and International Policy*, Arabic translation by K. Zakaria (Beirut, 1974) pp. 103 and 114.
24. H. O'Conner, *World Oil Crisis,* Arabic translation by O. Makkawi, (Cairo, 1967); also L. Mosley and B. Ratchov, *op. cit.*
25. I. Ahmad, *Rose el-Yusuf* weekly (Cairo, April, 1976) no. 26.

Part Two

1. H. O'Conner, *The Angolo-Iranian Dispute* (N.Y., 1962); also D. Wise & T. Ross, L. Mosley, Sampson *op. cit.*; R. Al-Barawi, *Oil War in the Middle East* (Cairo, 1953) in Arabic; M. Al-Mousawi, *Iran in Quarter of a Century* 1972, in Arabic; A. Sulaiman, *Arab Oil and Gas Review*, May, 1966. (Arabic).
2. M. Al-Mousawi, *op. cit.*, p.17.
3. *ibid.*, p.19.
4. *ibid.*, p.21.
5. O'Conner, *World Oil Crisis,* pp.361 and 2.
6. A.R. Muneef, *The Partnership Principle and Arab Oil Nationalization* (Beirut, 1973) p.60 (Arabic).
7. *Oil and Petroleum Year Book* 1951-1971.

## CHAPTER 3

# REBELLION... AND THE TEN-YEAR PUNISHMENT

Nuri As-Said himself (the pro-British ruler of Iraq) was complaining of the obstinacy of the companies. While he was in London to attend the Baghdad Pact meeting, he stated that his country had nothing to offer to foreign concession-seekers, because the IPC refused to yield any of the land they control. He also said: "Great anger erupts among the Iraqis when they look at agreements such as those concluded between the Japanese Company and Kuwait or Saudi Arabia. The IPC has to appreciate our situation".

But the company refused to appreciate it. On 4 July 1958 Mr G.H. Herridge (IPC Managing Director) arrived in Baghdad to start talks about concession agreements. The revolutionary feeling was growing in the entire Arab area after the nationalization of the Suez Canal, the political defeats of British and French imperialism, and after the defeat of the Eisenhower Doctrine, the representative of American imperialism with which the US aimed to inherit the area, and after the achievement of unity between Egypt and Syria. Iraq was influenced by all these explosive factors at the time of the arrival of the negotiating team. But the companies' representatives were so far removed from the Iraqi viewpoint that it was impossible even for the rulers of that time to accept their proposals. The companies refused to introduce any serious modifications to the concessionary agreements. Saying that the agreements were "holy", they refused to return any amount of land to the Iraqi government, because the provisions of those agreements did not give the Iraqi government that right.

So the negotiations failed, and the companies' delegation returned to London. Observers say that the delegation returned on 12 July full of confidence and optimism. Their calculations estimated that the Iraqi government would soon give in, and everything would remain unchanged. But, two days later, they received the news of the revolution.

The *New York Times* oil editor commented on this news by saying;

> "Concerning the Iraqi government, which used to be considered the most stable and advanced government in the Arab world, and the only government which the West has felt to be really dependable... this removal of the governmental system in Iraq may lead to serious economic and political consequences throughout the free world. It

may mean that it is not likely for any government, friendly with the West, in this part of the world, to remain in power for a long period. It may also weaken the not too tight western control of the vast oil resources in the Middle East".[1]

Several years after the 14 July 1958 revolution, Humphrey Trevelyan (who was appointed British Ambassador to Baghdad after the revolution) tried to evaluate what had happened. "The Hashimite-British alliance had outlived its time, while Nuri As-Said, the Prime Minister who became a stiff middle-aged man, careless about security affairs, failed to adapt himself to the political needs of the time... The confrontation arrived, and in one day all ancient ties with Iraq were broken... and will never be reclaimed."[2]

But, how did the imperialist policy of the Western States and the oil companies adapt to the new situation? Certainly they devised new plans on the basis that the removal of the monarchy in Iraq may have serious economic and political consequences throughout the free world but those plans did not 'adapt' to the political needs of the time. They took the other part of Trevelyan's evaluation, and built their calculations on the assumption that the question was basically one of security, which meant that they had to interfere in the policies of the producing countries, and that their security systems had to be more active and efficient, and, through control of wells in Kuwait, Iran and the Arab peninsula, Iraq could be sieged and its claims jeopardized. If insistent on revolt, then the Iraqi economy could be suffocated by stopping oil production and export. Compensation of consumers was easy through other producing countries... The same Iranian scenario.

In any case, what was the oil situation when power changed hands in Iraq in 1958? The oil industry in Iraq was organized according to four agreements, those of 1925, 1931, 1938 and 1952. It is natural that all these agreements represented the balance of power which was then prevalent. In other words, the concessionary provisions were formulated to suit the interests of the cartel companies. Basically, we can say that the general specifications of those concessions provided for the following:

1. The IPC, MPC (Mosul Petroleum Company) BPC (Basra Petroleum Company) should extract, sell and develop oil from the entire area of Iraq.

2. The companies did not submit, in planning or administration, to any participation of or supervision by the Iraqi state. The older

concessions had provided that the Iraqi government had the right to appoint one director in the Board of Directors of each of the companies. But this was amended in the February 1952 agreement when Iraq was given the right to appoint two directors. But the internal regulations of the companies provide for two different kinds of directors: there was the larger circle of directors representing each of the companies, where the two Iraqi directors belonged, and there was the narrower circle whose members were chosen from the first and wider circle – who were the group really in power – who planned and carried out the policy of the companies, and voted on the budget. Members of the narrower circle were called the executive directors and, among them, was the President of the Company and the Managing Director-General. The Iraqi government representatives were strictly denied entrance to this circle.

3. The concession period was 75 years (ending in the year 2013). The companies were careful not to provide in the concession for the need to relinquish any part of Iraqi land which the companies may control during any period of time. That is to say, the companies monopolized the right to exploit or freeze the entire area in Iraq until the end of the concession period.

4. With all these powers, it was natural for the returns due to the government, according to the agreements, to be the subject of constant fraudulence. When the oil companies were obliged to pay a royalty of 4-6 gold shillings for the ton until 1952, they used to evade even this meagre commitment. They did not pay the royalty on the basis of the price of gold in the free market, but on the basis of the official price, which was much lower. While Iraq was getting 330 fils, Saudi Arabia was getting 425 fils for each ton. When the division of profits agreement was concluded in 1952, the application was also subject to fraudulent interference in cost calculations, depreciation cost, payment of royalties, exploration and drilling expenses, open letting, loan interests, publicity expenses... naturally, this meddling with cost meant that the Iraqi share of profits was lower than the 50% specified in the Agreement.[3]

When the revolution broke out, the angry masses in the streets called for the nationalization of oil. But the leadership realized that nationalization needed preparation. Such preparation needed the modification of the whole system of relationships between the Iraqi state and the oil companies, so that oil should become a sector of the Iraqi economy not completely directed from London, far away from the sovereignty of the state. The calculations of the new government said that oil relations reflect the balance of power and the political situation. That power and situation had changed, therefore the change inevitably had to be seen in a new agreement, though it may not realize full nationalization and sovereignty. It would, however, form an important step towards the goal.

But the international oil cartel, backed by the imperialistic states, had a different idea in mind. As soon as they regained their breath, they began to look at Iraq as a small link in the chain of their branching international operations. If one link should break, then any leniency would lead to a kind of 'chain effect' leading to the explosion of the remaining links. And, using the remaining parts of the chain, which were still under their control, they decided to face the Iraqi revolt and kill the insurrection at birth. Therefore, the companies decided to regard this as a 'temporary' change in the balance of powers inside Iraq, and to translate this attitude into obstinate refusal of any leniency in dealing with the new regime. So, the two sides were diametrically opposed, and there was no hope of their meeting.

On 18 July, four days after the success of the revolution, it was officially announced in Baghdad that the government had taken necessary measures to protect oil fields and establishments, and the Government hoped that parties concerned in these areas would co-operate in order to develop this vital source of wealth. To put the statement in practice, the revolutionary government resorted to negotiations with the companies, hoping to reach a mutual understanding in the light of the new facts. In January 1959, the first long period of negotiations actually began. The formation of the companies' delegation was in itself a *tour de force* which attempted to show the unity of the international front which the delegates represented. Besides Mr. Herridge (IPC Managing Director) there was Stevens (representing Shell) and Fisher (Vice-President of Standard Oil of New Jersey Company). The main instruction given to this team was: refusal of any basic modifications in the 1952 agreement. The negotiations faltered in the first round, then the communications continued in various forms until April 1961. But the companies insisted on refusal of any real modification in their stand, and all international cartel companies supported the unrelenting attitude of IPC, resorting to the usual threat of proving their power by being able to replace the Iraqi oil by increasing the production of the other neighbouring countries. Quietly, instructions were given to local supervisors, and the annual decrease of Iraqi oil during the three years of negotiations was 5%-1%, while the oil production in Iran went up to 12% as an annual average during the same period; in Saudi Arabia the average of annual increases went up to 9%, in Kuwait to 11.5%. The Iraqi oil production in one critical month during the confrontation went down by 30%.

The demands of the Iraqi negotiator were principally as follows:[4]

1. The appointment of Iraqi directors, with real authority, in the management of the companies in London, and the supervision by the Iraqi government of company expenditure in a way that would secure Iraqi interests.

2. The necessity of the Iraqi government holding not less than 20% of the total capital.

3. Relinquishing of areas undeveloped by the companies in preparation for Iraqi development.

The companies did not believe that there had been such a change in the balance of power that they need make concessions over these questions so they considered them simply not open for discussion.

Negotiations failed and pressures continued. When the companies announced that they had stopped temporarily their operations at Rumaila, a field rich with possibilities, it was an occasion to issue the historic law, known as Law No. 80 of 1961 (published in *The Iraqi Gazette* on 12 December 1961). According to this law, the Iraqi state recovered 99.5% of land under control of foreign concessions, which the companies had not developed, and Iraq in turn had no chance to develop directly. This law meant:

1. That Iraq was turning from negotiation to recovering her rights by unilateral legislation.

2. That the companies, as a result, had lost their special status and had come under the laws of the state.

3. That the companies were being deprived of control of rich reserves, and the state was being provided with the opportunity of direct development of these reserves.

The companies had only 0.5% left in their hands of the area that had been controlled by concessionary agreements, that is, the entire area of Iraq. That 0.5% was not a little thing in practice. It meant according to the estimates at the time 25% of the area of where oil was expected in economically exploitable quantities. In oil potentialities, the Iraqi lands are not of equal value. Some areas are excellent, some merely good;

some accessible, others not; some rich in reserves, others poor. Like any other country, Iraq does not float on a sea of oil. There are geological conditions which led to the formation of oil and its accumulation in safe oil-traps. The areas with a capacity to receive 1,000 formations of oil do not exceed 50 thousand square kilometres, that is to say 11% of the total area of Iraq. No-one expected to discover actual oil traps in all this nominated area. According to the best expectations, the oil-containing formations were estimated to cover only 4% of the area of Iraq. The formations containing commercial quantities form 50% of oil-containing formations, i.e. 2% of the area of Iraq. When Law No. 80 left the companies 0.5% of the Iraqi lands of proven reserves, then it provided for the Iraqi state to operate within the remaining 1.5%. Dr Abdulla As-Sayyab says: "The principle of relinquishing land during the oil negotiations of 1961 was discussed on these scientifically sound bases. It was found that the area or the percentage were not so significant as the oil potentialities in the various geological formations. On this sound principle the final maps of Law No. 80 of 1961 were drawn".[5]

But the oil companies were not ready to listen to this or any other logic. Since they refused to recognize any political change affecting the balance of power, then what they refused to give to Nuri As-Said they had also to refuse to give to the new rulers, and with more obstinacy. Consequently the companies denied the legality of Law No. 80, of which they knew before its publication at the conclusion of the 1961 negotiations. The official minutes record that it was said to the companies' delegation: "These are your wells which you can exploit as you wish. We are sorry to tell you that we shall take the rest of the lands in accordance with our prepared legislation, so that you will not be surprised". The chief delegate, Mr. Fisher, answered very confidently: "We thank you for telling us of that, and we shall see what the results will be".

I think that any historian of this period can safely say that Fisher was right in commenting so confidently. The Law No. 80 was an important and courageous step since the days of Musaddaq, and for several years that followed. The individualistic rule of Abdul-Kareem Qassim was really not able to face the consequences resulting from such law and fighting among patriotic and progressive forces had broken down everybody's power to struggle against a common enemy. Such fight among patriotic forces did not only divide the ranks of the Iraqi masses, it was devastating to the Arab world as a whole. Undoubtedly

Fisher had this image in mind when he gave his open warning: "We shall see what the results will be." He realized that in the end he owned the producing wells, and since the Iraqi government was politically weak, and since it relied on money given by the companies, then the companies could force the government to freeze the law, if not to withdraw it completely.

Fisher gave his warning, and that was followed by the freezing of Iraqi production, which, in 1962, did not increase by more than 0.3%. It is true that the Iraqi government took another step after Law No. 80 and drafted in 1962 a law to form the Iraqi National Oil Company, whose draft gave this company "the right to exploit all areas containing oil and natural hydrocarbon wealth in Iraq, except what is specified by the law limiting areas of exploitation by oil companies, No. 80 of 1961". But it was difficult in circumstances prevailing at that time for the matter to go beyond the stage of talking.

But freezing the situation was not a solution. Another round of negotiations started after the end of the individualistic rule of Abdul-Kareem Qassim and lasted from 5 February 1964 to 3 June 1965 and in a strictly confidential atmosphere. The result was a report and outlines for two agreements, the first to amend the 1952 concession, the second to form the Baghdad Oil Company.

According to these three documents, the lands with oil potentialities were returned to the companies. These lands included: North Rumaila Field, Tawiya Field, the northern extension of Zubair Field (in the southern area). And, in the northern area, it included South Kirkuk Field, South Bay Hassan, North Jamhour and Ain Zala. The area of these lands totalled 1937 square kilometres and held 3,000 million tons of proven reserves. This relinquishing of lands undoubtedly meant a cancellation of the basic content of Law No. 80. The mere fact of relinquishing these fields implied that the companies had a right to these regions. On the other hand, what was left to Iraq after this retreat was the oil of the lands with doubtful potentialities.

It was natural after this that the documents which came out of the negotiations should kill the other result of Law No. 80, which was the direct national exploitation of Iraqi oil. The "Baghdad Oil Company" was to have seen to this job. This company was supposed to be a joint enterprise between INOC, paying one third of the capital (in return for relinquishing 33,000 square kilometres) and the cartel companies forming IPC, paying two-thirds.

The Iraqi delegation to the negotiations justified these and other essential concessions as follows: "It became clear that the difficulties which beset the oil industry in Iraq, from the issuance of Law No. 80 to the present, cannot be removed without the companies submission to the law, and the application of such law on the companies with their consent". Concerning the relinquishing of the independent role of INOC it was said: "It was found that the mere formation of the national company was not enough to achieve this aim [exercising oil operations], since it is also necessary to solve the problem of the companies operating in Iraq." Then, "the national company is unable to invest tens of millions of dinars in oil production in those lands, and is unable to find markets to sell that oil, because the company is naturally unable to fight the international monopolies controlling these markets, and is unable to conclude contracts with foreign companies to ensure marketing, since such companies do not conclude such contracts for oil which is a subject of dispute between the company and the international monopoly companies, thus risking their future".[7]

I have chosen to publish excerpts of this report because they remind me of the same justifications that were advanced to oppose nationalization after the July 1968 revolution. In fact the reasons which deny nationalization are at the same time reasons which mean surrender to the companies. But the resistance of patriotic and progressive forces did away with these projects which were based on open collusion. Tahir Yahya's cabinet resigned, and matters developed in a radical direction. Law No. 97 (6 August 1967) was issued to emphasize the basic principles of Law No. 80, the role and area of INOC. The policy of oil development of this company was defined as being by direct exploitation of national oil-fields or through cooperation with another party. The law was, as commented Dr Muhammad Salman Hassan, a step forward on the path of oil liberation in principle, but its real value depended on sound application.[8] This sound application was not possible without a political authority capable of challenging the monopolies which had been refusing to recognize the content of Law No. 80, and the right of Iraq to produce oil from the areas which the law had salvaged. The companies were quite sure of their capabalities to challenge. When a high official of that time said to the companies' delegates that the companies' arrogance might force Iraq to nationalize, the answer was: "Please do! "

The companies were sure Iraq would not do so. Pressure continued in one way or another for "taming" and securing concessions from the government. In the period between signing the 1952 agreement and the end of 1961, the average of annual growth in Iraqi oil production was about 11.1% But the end of 1961 witnessed the issuance of Law No. 80, and the average rate of annual growth from 1962 to 1970 was 4.7%. This punishment cost the Iraqi economy a loss of income of about 550 million Iraqi dinars. It cannot be other than punishment. The annual average of increase in Middle Eastern oil was 13% in the same period of 1962-70. The average increase in oil consumption in the world was 9% a year. The production cost of Iraqi oil is almost the lowest in the world, but all these favourable economic considerations were countered by the companies with the claim that Iraq was a trouble-maker and always complaining. And that is a political consideration... and in the question of oil, politics comes first.

Incidentally, this discussion can no longer be considered either prejudiced or guesswork. With the passing of time, the oilmasters disclosed the secrets of their game. Anthony Sampson wrote: "The Iraqi indignation at the companies became more acute after the 1958 revolution. Iraq helped to push other Arabs to take radical stands, and helped in the birth of OPEC. But Iraqi oil production relapsed during that time, though Iraqi reserves were larger than those of Iran." The writer adds that the existence of various resources of oil production in the hands of giant companies always enabled them to manoeuvre with freedom. He quotes Howard Badge as saying that Middle Eastern oil production increased so rapidly by virtue of decrease in growthrates of the Iraqi product. The relationship between radical stands by oil producers and decrease of their oil production was not left unanswered by Badge. Any country that creates many difficulties for the companies provides a good reason to decrease its production. In Howard Badge's own words: "Sometimes they make it easier for us to decide on the decrease: they break the agreement – as happened in Iraq – then we can tell them: Go to Hell." In quoting Howard Badge, Sampson comments by saying that "Iraq became a kind of new warning to the others, exactly as Iran was under the rule of Musaddaq".[9]

It so happened that the Shah was insisting to increasing production rates in Iran during 1966. The Iranian authorities felt the companies favoured Saudi Arabia, and Howard Badge was busy in communications with the two states. When the Shah complained to the US State Department, Jim Aikens (State Department oil expert) told the Shah frankly that he could not agree with the Iranian point of view. Equally frankly he said: "If the Iranians become very provocative, they may share the destiny of Iraq".[10]

The matter is no longer a secret, then. But it may be asked: who is this Badge? But to introduce the man it is necessary to point out that the seven (or eight) cartel companies are all linked by joint projects. In the midst of the circle sits the American Exxon Company (or Esso). This company is a partner in Aramco, the Iranian Consortium, and IPC. The circle master who led the negotiations in the Middle East was the Business Manager of Esso who has been for the last twenty years: Mr Howard Badge.

**NOTES**

1. H. O'Conner, *op. cit.*, p.383.
2. H. Trevelyan, *The Middle East in Revolution.*
3. M.S. Hassan, *op. cit.*, ch.4.
4. *Vide* Official Statement by the Ministry of Oil, April 10, 1961.
5. A.S. As-Sayyab, in *Arab Oil and Gas Review,* February, 1967.
6. Statement by Iraqi Ministry of Oil, October 17, 1961.
7. Report by the delegation for negotiations with oil companies, headed by A.A. Al-Wattari in the negotiations of May, 1964.
8. M. S. Hassan, *op. cit.*, pp. 94 and 95.
9. Sampson, *op. cit.*, p.167.
10. *ibid.*, pp. 172 and 173.

## CHAPTER 4

# ENCIRCLING THE FORTRESSES

At 6 p.m. on 30 July 1968 the Revolutionary Command Council published a statement about "the liquidation of the faction conspiring against the Party and Revolution – the Naif and Dawood faction". This incident meant that the leadership of the state passed fully into the hands of the ABSP from 30 July, 1968. Therefore, it was natural that the RCC statement should include the outline policy of the new government, and it was inevitable that this policy should touch upon the question of oil. So, the statement spoke about "extracting the legitimate rights of Iraq in its oil wealth, and removing all the injustice done to Iraq as a result of negligence on the part of the previous reactionary governments, and their conspiracy against the people's right to claim the benefits of the country's wealth in their entirety, and to develop its oil in a manner which would be fitting to its sovereignty and yield to the people what would further its progress".

No doubt these words were chosen with great care. They are a commitment to the general aims of struggle. But they deliberately avoid any exact limitation. How are rights extracted? And what are the legitimate rights to begin with?

All these questions were meant to be left open. The new Revolutionary government was hardly settled, and there was no sense in putting its cards on the table with reference to its intentions, which could hasten confrontation when the time for it was not yet ripe. On the other hand, the Party Command, after reaching power, was in a completely different situation. The Party came to power, this second time, having learnt from past experiences. And, according to these experiences, the Party was conscious that oil nationalization could not be achieved without a comprehensive revolution, necessarily demanding preparation and various stages. The masses in Iraq as in all the Arab countries were living through the years of fatal disappointment, after the June defeat. The Iraqi and Arab masses were not in need of resounding mottoes for chewing over, or great speeches just for listening to and nationalization no longer needed to be extolled. What was needed of the ideology that claimed to be revolutionary and progressive – especially when it was backed by power – was to be carried out!

It was inevitable then that the Party Command start immediate application. Even the study of the stages and transitionary steps towards nationalization did not need much hard work. The new government needed to start from the point where the previous course of struggle had ended. That struggle had reached the stage of sound theoretical planning. A national attack to liberate oil wealth could take two directions: The first was to start a skirmish with the cartel companies which owned the concessions, and to use the utmost pressure possible to check their authority and exploitation. The second direction of attack would be by commencing oil production by national teams on the sites recovered by the state through the Law No. 80. There is no doubt that these two branches of attack complement each other. The first mode of attack would open wider horizons for direct national oil development; because as long as the concessionary companies maintained their great power inside the political and economic system of Iraq, and as long as they maintained their influential international relations, the national production and marketing of oil would be a more difficult job. At the same time, developing the attack in the second direction would train the Iraqi staff in the methods of planning and marketing, which the monopolies had hitherto preserved with secrecy and an alarming atmosphere of mystery. Moreover, earning oil royalties through national development would deflate the feeling of the concessionary companies that they alone had the keys to the Iraqi treasury. The plan for direct development would help in surrounding the "major forts" to be stormed and weaken their challenge. In the end, the two modes of attack meet at one goal: the complete liberation of oil wealth.

This was the plan, and the already established Law No. 80 was an attempt to attack and undermine. The plan was sound and needed no revision. But what needed revision from any serious leadership was the practical calculations now that the relevant data was available to the new government... then the next step was to start action immediately. And this in particular was the difficult thing, just like the old story of the plan to hang the bell round the cat's neck to alarm the mice. The plan is pretty. But who shall hang the bell, and how? That was the question.

The companies realized, of course, the intention of the Party Command to ask them for partnership in ownership and administration. This was a change in the system of relations, which, when realized, would certainly be followed – under a revolutionary rule – by a full nationalization. The companies also realized the Command's intention to demand the return of some of the lands in their control to be developed nationally, and may well have realized that was the second mode of attack which would soon uproot them. This was an intention published and announced by the national and progressive forces. The companies realized that a surrender in A would lead to a surrender in B which, in turn, would lead to C etc... Therefore, the companies refused to move from their positions which ensured full control, and consequently refused Law No. 80, which they took to be A or the introduction to the comprehensive change. The attitude of refusal was understandable, and the companies were conscious of the dimensions of the game, and the game was played with open cards on both sides. Refusal was also understandable because the companies thought that the balance of power could still make them impose a continuation of the *status quo.* They had really succeeded in freezing Law No. 80 and in scheming round its contents for years.

But a new government had come to power, and the seriousness of its intentions in the area of oil would be tested by their courage in the practical implementation of the nationalization plan. The Ba'th leadership began by accepting the challenge, and decided that basic attack in the first stage should be towards national development, on the assumption that real advance in this direction emphasized the legitimacy of Law No. 80, and proved that the new system of government was different from previous systems, so by the time the skirmish with the "major forts" came it would be from a stronger position.

This was the decision, and action began immediately. As soon as the 17-30 July revolution settled down, and after the decision to make a concentrated approach to direct national development, came the approval of administrative formations and internal regulations of INOC. After that, the company Board of Directors outlined the oil policy. Within this framework a technical team laid the bases of the first ten-year plan.

After this, it became necessary to provide financial and technical resources. The monopolizing companies had made large efforts, using their influence inside the international cartel, to stop various parties in the world from co-operating with INOC. Those companies created obstacles against the formation of combined foreign and Iraqi companies for geophysical services, exploration and drilling, and used pressures and threats against western suppliers to force them not to provide INOC with oil tools and equipment. The monopolizing companies succeeded at first in influencing some foreign parties which were ready to co-operate with INOC in land exploitation. Those parties made poor offers which did not compare to current agreements concluded with neighbouring producing countries. The reason given was that those parties were taking large risks by working in Iraq and thus becoming vulnerable to monopolizing companies abroad.[1]

It was necessary to face up to the difficulties. The rise of a national oil industry,

> "demands advanced technical standards, huge financial resources and large marketing possibilities not within the reach of Iraq as a developing country. To meet these important needs, in addition to execution of various other projects included in the country's development plan, makes it necessary to deal with foreign parties of large financial abilities and high technical efficiency. Yet the basic question remains to find the parties which can co-operate with Iraq honestly, and this simply means we must stop dealing with institutions of an exploitative nature, and to approach the parties which deal on different bases, characterized by co-operation and mutual economic interests.Hence the change in our oil connections towards friendly Socialist countries, where good will and plain dealing free from exploitation prevail".

. . .

In accordance with this concept, numerous steps were taken: a Follow-up Committee on Oil Affairs and Implementation of Agreements was formed (six members of the Party Command headed by Comrade Saddam Hussein, aided by a technical secretariat). The Committee – as the title indicates – was entrusted with powers of action to implement the political course adopted by the Command in the field of oil. In the same year (1969) important agreements were concluded:

21 June 1969. An agreement was concluded between INOC and Machino-Export. The Soviet establishment agreed to provide the Iraqi company with all equipment and technical expertise necessary to build up a national oil industry. This agreement broke the technological siege exercised by oil monopolies, and also helped in solving the financing problem. The agreement provided that Machino-Export should offer INOC a 25 million Iraqi dinar loan, at a small interest, to be used in importing equipment, needed for the Iraqi company operations, from the Soviet establishment.

4 July 1969. A new agreement was concluded emphasizing the same

approach. The Iraqi government concluded an agreement with the USSR government for economic and technical co-operation in developing the oil production industry. In this agreement, Iraq obtained a loan of 24 million Iraqi dinars at 2.5% interest, counted on the part used from the loan. This amount (or a larger amount as the agreement provided for the possibility of its increase) was to be used to meet the expenses of Soviet establishments carrying out some of the projects necessary for the development of a national oil industry. The payment of these loans was not difficult for the Iraqi economy, since it was made by crude oil shipments which the Soviet equipment helped produce from INOC fields.

21 October 1969. INOC concluded an agreement with the Hungarian Kemocomplex to drill four wells in North Rumaila Field. The agreement provided that the Hungarian side was to offer a 15 million dollar loan to buy material and equipment, and to cover services offered by the Hungarians. The Hungarian side expressed readiness to buy crude oil produced from this when shipment was made possible.

I hope the reader will recognize the serious significance of these agreements of 1969. They are not merely financial and technical aids extended by socialist countries to a developing country. Even in that sense, the agreements were annoying to western countries. But the degree of annoyance became more acute because the developing country was also an oil-producing country, and the assistance was in the field of oil in particular. Oil is a commodity of prime strategic importance, and western and American policies endeavour to monopolize the technology and administration of this industry. To break this monopoly would cause a serious threat to the possibility of continued control of oil production sources. Moreover, the aid was for developing North Rumaila Field, an area of conflict between cartel companies and the Iraqi state. Therefore, the aid in this case cannot be considered of merely a technical and financial nature: it was also political support to help turn Law No. 80 into a tangible reality, and consequently support for the stand of a country fighting monopolies to liberate national oil wealth.

Tariq Aziz was not exaggerating when he said: “If a comparison were useful, then the agreement with the Soviet Union to develop North Rumaila Field nationally is equal in historic importance to the Egyptian step in 1955 to obtain arms from Socialist countries.”[2] Joe Stork wrote: “The INOC agreement with the Soviet Union was described by experts as the most important development in oil industry in modern history".[3]

It was natural that the wheel should take a complementary turn inside Iraq. In June 1969, INOC took a step outside Baghdad and set up a branch under the name of "The Directorate-General of INOC in Basra". There was no concern about appearances; the modest headquarters occupied a school building, from where actual preparations were started for work in North Rumaila Field. On 15 July 1969 the corner-stone was laid for the development project of the field. Early in 1970 some of the drilling equipment arrived on the site. And in February 1970, Law No. 24 was issued to fill a legislative gap in Law No. 80. Law No. 80 had a notable article (No. 3) which specified that "The Government of the Republic of Iraq may allocate other areas to be as reserves for the companies, provided such areas do not exceed the area allotted to each company." This article was a legislative Achilles' heel in Law No. 80. It left a backdoor open for the companies to be able to get to the most important areas they used to control, given favourable political circumstances. There is no doubt that cancelling that third article was another indication that the revolution was not ready to bargain about Law No. 80.

But the companies had not received notice of these events, or they had not understood them, when preparation for a confidential round of talks got under way.

The President warned, on the second anniversary of the revolution, 17 July 1970, that "We declare that the Revolutionary Government will spare no effort to secure the rights of Iraq, and shall not accept any postponement, procrastination or side-stepping of these rights". He added: "Iraq cannot accept impossible terms and outdated claims in return for its natural right to royalty expenses. Such matters of vital importance for the interests of our nation can no longer be left to time. The Revolutionary Government is determined to deal with such matters and take all necessary measures to secure our legitimate rights".

At that time, the President did not announce the negotiations with the companies. But we can now understand the reason behind the sharp tone of this warning. It seemed that the companies were still ignoring what was happening around them. They refused again to show any degree of flexibility, even in matters that sound simple, like the question of royalty expenses which was decided in all oil countries. The companies insisted on relating their approval on royalty expenses to basic concessions on the part of Iraq which would lead to the abolition of Law No. 80. The basic issue raised by the Iraqi delegation in these negotiations concerned the agreement on reasonable and stable rates of production increase. That was completely refused, and the most that the companies "condescended" to offer in September 1970 was some payment of royalty expenses, to be deducted from Iraqi dues at the final settlement of the issue, in addition to production increase in the southern fields to an average of 28 million tons.

Naturally, it was not possible for a revolutionary government to accept such a proposal. The companies' insistence on scheming and side-stepping Law No. 80 meant that the companies were still insisting on the old system of relations which did not allow for any sovereignty of the state over its oil industry. Consequently, they insisted that negotiations should remain within the existing system in return for some partial improvements. And even within the framework of partial improvements, which increased the state royalties, the companies showed a curious arrogance.

The negotiations were in fact a kind of reconaissance skirmish. When the Party Command held a meeting to discuss the results of that round, it was concluded that the companies were still insisting on confrontation, and that they had not even resigned themselves to the idea of retreating one step, with a hope of containment after that. Consequently, the appraisal was that the possibility of coexistence between the national government and the companies on the soil of Iraq seemed out of the question for some time to come. The decision was issued to prepare for a large confrontation. The Follow-up Committee had to prepare itself, and this was reflected in accelerating action towards national development, and in confidential discussions and debates about nationalization.

Undoubtedly, the companies did their part in evaluating the situation in the light of the failure of the reconaissance round. Most probably they had noted the firm Iraqi attitude, but their assessment was still that it was possible to check the advance of the enemy towards any "radical" aims, and if not, the punitive measures were ready.[4]

## NOTES

1. INOC Paper presented to International Seminar on Oil, Baghdad, 1972.
2. T. Aziz, *The New Course Revolution*, (Baghdad, 1974) p.29 (Arabic).
3. J. Stork, *op. cit.*, p.89.

4. All information in this chapter comes from official documents and publications by the Iraqi Government, or from conversation between the author and political and technical personalities.

* Royalty, in the production industry, is a term used to describe the income received by the owner of the natural source, in his capacity as owner. There is no difference when the owner is a state (as in the case of Iraq) or an individual (as in the US) since in both cases the owner gets the royalty. It is supposed that this income is not connected to taxes levied by the state on profits of oil production and exploitation operations. But this separation between royalty and tax was not followed in the dealing of the companies with OPEC countries for a long time, which resulted in injustice in calculating the share of profits due to the oil-producing country. Iraq, for instance, used to get 50%of profits, but the royalties were included in this proportion. Paying royalty expenses means including the proportion given to the country in return for use of their resource (which was 12.5% in the case of Iraq) within production expenses, as something separate from the 50% of profits payable. This last payment is the tax levied by the state from the company profit. The profit equals price minus production expenses (including royalty). From this profit the 50%proportion is taken.

# CHAPTER 5

# ESCALATION OF ATTACK ON THE SOUTHERN FRONT

When drilling began on the first well at Nukhaila (North Rumaila, 15 July 1970), the Iraqis were working with unequalled enthusiasm. But when drilling began on the second well (23 September 1970) the Iraqi staff was running the operation completely by using the drilling rig. The local people completely succeeded in their assignment, and the well was ready for production by the end of February 1971.

Before my visit to the southern fields, I used to think there was a dividing line between North and South Rumaila Fields. But on the site, you do not need the geologist's explanation to realize that the line was imaginary. North and South were two branches of one huge field. INOC was boring the ground in the north, and Basra Petroleum Company (BPC) was producing from the same field to the south.

When you go from Basra north-east to Rumaila region (60 kilometres) you first reach the work sites of North Rumaila. But when your car continues, you will be told, though you do not notice it, that you are now in South Rumaila Field. In fact INOC did not need to drill exploration wells[1] nor was there a need for evaluation wells.[2] The first and second wells, and all the wells of the first stage, were developmental or producing wells. The drilling process of this kind is cheaper and easier. The INOC technical departments assigned ten wells to ensure the projected preliminary production stage (5 million tons per year). At the same time drilling had to be completed in the shortest time possible. INOC gave quick training courses to a large number of Iraqi staff, and concluded agreements with Arab staff, (the Egyptian drilling staff enjoyed an international reputation) and Indians, too. By 15 September 1971, the ten wells were ready for production.

But the question was not simply a programme of drilling a number of wells. Drilling had to be within an integrated plan (and this was the first lesson the national staff had to learn). With the completion of the drilling programme of the first stage, it was necessary to make studies of field develpment, and to plan product enlargement for the second stage. This had also to be accompanied by balanced programmes for a network of projects necessary for production, transport and loading, and also for ancillary services.

After deciding the types and sizes of projects, it was necessary to make agreements for the import of supplies, and to provide co-ordination between dates of arrival and dates of completion of civil and local constructions to safeguard against delay in installation of equipment.

A large and complicated operation – much more complex than the previous paragraph indicates. But, the responsibility had to be faced, and learning by practice was the only solution, and, naturally, there was nothing against using outside experience.

Yet, in accordance with plan, the ten wells for the first stage production were ready by 15 September. When the drilling was over, the experts had finished their job. Degassing stations, pumping stations, and crude oil reservoirs were set up in Rumaila and Fao (South of Basra). The pipeline to carry crude oil from North Rumaila Field to Fao port was also completed. With the co-operation of the Iraqi Ports Administration, the company prepared Fao port to receive the 35 thousand ton oil-tankers.

But there was no time to rest. At the same time as working on achievements it was necessary to prepare for the second stage. INOC contracted with Techno-Exporting in May, 1971 to prepare consulative studies on development and preparation, capacity was to go from 5 to 18 million tons. Rumaila development was part of the ten-year plan for oil sector development, which was prepared at the time with the cost estimated at 498 million dinars. The plan covered operations of geological and seismic surveys; exploring, developing and producing drilling-projects; transport projects (pipe-lines, tankers and ports); civil engineering and laboratory building projects; and natural gas development projects.

In the midst of all this, the National Action Pact announced by the Ba'th Command in November 1971 said under the heading: "Oil and Minerals Policy":

> "At this stage, oil forms the basic revenue in the economy of the country, and plays a decisive role in defining the country's economic and political future. And because of this exceptionally important role of oil in our life, the foreign monopolies have been playing a serious part in threatening the independence of the country, and conspiring against the progressive and national forces, and against the people's aspiration to achieve liberty and progress.
>
> "It is natural that the strategic aim of the Revolution should be the complete liberation of oil wealth from foreign control and exploitation, and to subject it to national sovereignty in every way,

and to use it fully in the service of the people's welfare, and to make it a decisive tool of struggle against imperialism and Zionism.

"The scientific and serious work to achieve these aims demands a continuation of our previous stand in the course of the oil policy adopted by the Revolution, which aims in the first place to set up a large, strong and integrated oil industry.

"The building of a national oil industry demands the support of the industry with all material and human possibilities at its disposal and a development of institutions and establishments which provide the staff needed in the oil industry, as well as co-operation with the friendly Socialist countries."

We notice that the discussion of oil policy is more defined this time than it was the case with the 30 July 1969 statement. This clearer definition was not due to an increase in theoretical understanding, but was an expression that a stage had been actually completed. The Pact could not ignore the power of the companies, but it was preparing for a confrontation. Yet they realized that the time for confrontation had not come yet, and preparations for it were not complete; therefore the wording avoided the use of the "awful" word of nationalization, but at the same time recorded the content of the word by talking about the liberation of oil wealth and subjecting it to national sovereignty. The policy was at that time to concentrate the attack in the direction of direct development, and the Pact says this clearly and exactly. It is an example of the responsible use of words, without exaggeration or bargaining: words in line with the situation and its future possibilities.

Back to the south, with the men working in the oilfields and their friends. The work was hard, and it took years in the arid sands and the marsehs.[3] The basic comforts of life were not available. These are exceedingly backward regions. Some of the inhabitants knew nothing of what was going on in the world. Therefore, it was not easy for the area to witness all this activity all of a sudden, and to receive this number of new arrivals.

It was not possible for the work to wait until the services were set up. The work had to start and for the many, work used to start and never end.

It was natural that the development efforts should be faced with all the obstacles and difficulties that are caused by a backward social and economic system. This may have caused some delay in the rate of achievement on some sites, which in turn affected the other links in the chain, thus increasing the costs.

There is no sense in exaggerating when surveying these phenomena. This is a price that has to be paid; it is the price of learning and progress. Yet this should not excuse delays in trying to improve performance, otherwise, what is the sense of learning and progress? What is meant is simply that we should not be annoyed by these phenomena. The Iraqi experiment to develop the south, especially in the field of oil, was wonderful by any standard. The under-development problems were faced without hesitation, and whoever visits the INOC sites now cannot help but be astonished, not just by establishments and equipment working smoothly, but by the new type of man directing the work. There were 350 men working for the company when it started in 1968. The number jumped to 2,284 only a year and a half after the company moved to Basra, that is by the end of 1971.

Yet, sometimes the problems of underdevelopment, which confronted the men, were of a unique type.

Until 1971, there were two seismic teams to survey the Iraqi lands (an undertaking of vital importance which studies the geological formations under the ground's surface). The first was a land team, the other amphibious.

This second team, all Iraqi, (engineers and workmen) was divided into six sub-groups: survey, drilling, logging, electricity, mechanic, administrative and services. This team went to start work in 1971 in the area north of Rumaila to the west of Qurna, called the Zubair region.

To the west of Qurna, their work took place in the midst of the marshes, which was this team's speciality. Reaching the work-sites was only possible by motor-boats or canoes. The team began by planting mines at various distances and depths over the land. The mines were connected to an explosive device, which, when the signal was given, would be triggered.[4]

Before the explosion, the electrical engineer gave his usual order to stop all movement at a 100 metre radius to avoid harm from explosion vibrations. Then he phoned the man in charge of the explosion device to tell him to prime it, then he pushed the button... The men held their breath and waited... Nothing happened.. No explosion was heard.

It was an embarrassing situation. What added to the embarrassment was that the matter was not caused by any technical mistake. The cause was subversion. The inhabitants of the area were not pleased to see those outsiders in their region. Subversion is an easy job. One hit of a sickle can cut the wires, and the team had to search for days to find that cut. Also, it is easy to let the cattle loose to swallow some plastic geophones which were fixed on the surface of the ground or buried in it, and the work would stop again. Men have lived in this wild area for centuries, feeding mostly on fish and birds, and were suspicous of these people and their strange activities.

At the start, it was suggested the police be used. But this decision was overruled as it would have complicated things, especially because those people were armed, and they had often threatened the men of the second team. Some volunteered to talk things over with those inhabitants, but it was found that logic was not recognizable. The strangest thing is what one old man said to the team members: "We know who you are... You are *gendarmerie* coming to clothe us in khaki and send us to Turkey! " Was this imaginable? It had not reached them yet that the Ottoman rule had gone out more than half a century ago!

The solution was to re-educate these people through work, and through convincing them by personal practice that the new comers were friends who aimed to do good.

The health team started to extend its services to the inhabitants of the area. The food portions were doubled for each workman so he could feed his neighbours too. Then it was decided to hire the local canoes owned by the inhabitants for travel in the area, and they were reimbursed generously. A canoe that could take four people was hired for one person only. This cost the company an additional 20-30 dinars, yet it saved a possible loss of 800 to 900 dinars.

In this manner, confidence was established. The inhabitants believed that the INOC representatives meant them no harm. Through that confidence, more steps were taken to convince the able among the inhabitants to do some productive work. Many joined the army of INOC workers, which was an economic saving as it saved the company the trouble of bringing men from distant areas. But the greater gain was that the country recovered those citizens.

And how changed they became in the process of construction, and through dealing with tools, cinema and radio! They began as a problem, but, like other problems, it was solved through a programme of development of awareness.

With all these difficulties, everything was ready for operation in time. On 7 April 1972, Saddam Hussein inaugurated the operation of North Rumaila Field.

The Iraqi State, while fighting on Iraqi territory, did not forget the international oil front. Iraq actively participated in the Tehran Agreement in February 1971, which represented at that time a significant development in the life of OPEC. Then Iraq became increasingly involved in the later efforts which led to the first Geneva Agreement in January 1972.

Did the monopolies realize – after all this determination – the need to revise their calculations? We shall see. But it is worth saying here that the inauguration of North Rumaila Field was done while negotiations with the companies were gasping for life.

Before we carry on, it is worth mentioning the shipment of Rumaila oil. We have followed the oil from the field to the port. But the oil had to be loaded into the tankers which would take it to the markets of the outside world.

Providing tankers is a thorny question in the field of oil, as is also the organization of relationships with markets. These points did not escape the men who planned the policy of direct national development. In the area of shipping, contracts were concluded in 1970 to build seven tankers of sizes suitable for the capacity of Fao Port (35 thousand tons). INOC received the first tanker *Rumaila*, in mid-February 1972. This tanker enjoyed the honour of participating in the first contract made by INOC and a foreign establishment. As soon as she was received, *Rumaila* made her first voyage to Arzew port in Algeria, and was loaded with Algerian crude oil which she took to Rio de Janeiro, for the Brazilian Government Oil Company (Petrobras). *Rumaila* was the first Iraqi tanker to carry national oil.

## NOTES

Information in this chapter comes from official documents and conversation with technicians in Baghdad and Basra.

1. Exploration wells form the most complicated drilling operations to collect geological information: The aim of drilling such wells is to study the area geologically, and to study the layers of carbohydrate contents if found and to gauge pressure and thickness of oil structures, in addition to the type and physical and chemical qualities of the oil. Such drilling programmes are initiated in the light of information taken from seismic surveys and comparing data from neighbouring areas.
2. Evaluation wells are drilled when the existence of a hydrocarbon reservoir is assured. The aim is to evaluate the quality and reserves of such a reservoir and to get additional data on exploration wells. In accordance with such data, producing and development wells are located; the distance and number of such wells are decided. These wells are usually less complicated and costly than exploration wells.
3. The marshes are depressions in southern Iraq, filled with the waters from the

Tigris and the Euphrates. They form a difficult obstacle for development projects and search and exploration operations for oil.

4. The idea is that sound waves caused by an explosion are reflected in various layers differently, each layer of the ground reflecting these waves in a particular way. Then the geophones pick up these reflected waves and transmit them via wires to the logging station. At the station this "message" is recorded on a tape, which is sent to the company, where the geophysical team interprets and analyses this information.

## CHAPTER 6

# THE HIGH COMMISSIONER AND THE PASHA

It might be interesting to wait a few moments before moving on to the last round of negotiation, to recall some memories which exemplify the balance of power over the years and how it changed.

In 1924, the first communications and talks began between an Iraqi Government and a representative of the Turkish Petroleum Company (which later became IPC). In December of that year, the situation had crystallized, and it became necessary to concentrate efforts to decide the controversy.

Seven meetings were held between 10 and 21 December, at the residence of the British High Commissioner. H.E. Yassen Al-Hashimi (Prime Minister of Iraq) was present in the newly formed negotiation committee, together with Muzahim Al-Pachachi (Minister of Communications and Works) and Rasheed Ali Al-Gailani (Minister of Justice). On the other side was the representative of the Turkish Company, Mr Kittling, in addition to Colonel Tanish and Mr Swan. At the head of the negotiating table was H.E. the British High Commissioner, Sir Henry Dobbs, occupying his natural position as chairman and patron.

This arrangement alone is enough to summarize the political situation, and the balance of powers which controlled the negotiations. Yet, it is necessary to record that the negotiations were not easy at all. The Iraqi side raised all the basic issues, and defended the state's rights to sovereignty with understanding and firmness. But the representatives of the imperialistic powers – companies and governments – were not ready for any leniency; they were confident they possessed all the means of pressure which could enable them to impose all their basic terms, and even most of their legal terms.

The matter for concern now is the basic issues raised by the Iraqi negotiators. They had received a letter from the Ministry of Colonies No. B-O-266, dated 24 December 1924, addressed to H.E. The Prime Minister.

> "1. The British Government regrets that she cannot enlarge the assurances which she has previously given concerning legal and economic concessions. These assurances were included in two statements submitted to H.M. King Faisal on 12 February 1924. Your Excellency will see that the British Government had at that

time gone to the furthest limit compatible with the diplomatic style in promising to make her utmost effort to stop the re-enforcement of concessions. *And I feel confident that if the Iraqi Government followed the policy she is following now, there would be no danger whatsoever of the re-enforcement of these concessions."*
(These, and the following italics are by the author).

This is a hidden warning that a change in the negligent policy may lead back to the enforcement of concessions. In any case, the letter from the Ministry for the Colonies continues:

"Therefore I believe that the fear of the re-enforcement of concessions should not be given any weight by the Iraqi Government, when discussing the subject of the by-companies or the foreign companies, in accordance with the draft agreement [that is, to register those companies in Great Britain, and not as Iraqi Companies].... and I have understood from the British Government that *there is no hope the Turkish Oil Company will accept any alteration in the draft in this connection."*

Naturally, this is a matter of principle, denoting from the start the extent of the company's authority and the extent to which they intended to submit to the Iraqi Government.

"The British Government has made it clear that a very great amount of the shares must be bought by foreign concerns, and these amounts cannot be obtained unless the investors could be sure that the company they are investing in will, in case of need, be safe from danger, by registering it as a British Company. *I am confident that it would be possible for the Iraqi Government after these explanations to accept the proposals advanced in the concession draft concerning the registration of companies.*

"2. The Government considers that if the Turkish Oil Company, in lieu of the foreign or by-companies, should undertake not to profit by any economic privileges that may develop in the future, this contract shall hold valid. *But it is not likely that the company should give such an undertaking.*

"3. The British Government clarified that the right of the Iraqi Government, or local concerns, to participate in up to 20% of the capital, in accordance with the San Remo[1] agreement, is recognizable. Yet, the Turkish Oil Company has explained to the Minstry of Colonies that when the company agreed to pay royalties equal to four shillings for each ton, *the company had supposed that this payment would have excluded the idea of Iraqi participation in the capital shares.* Expenses of oil production and pumping to the sea are very high, therfefore, the share-holders will not get high profits from their shares in the Turkish Oil Company... So, the profits effected from selling crude oil will be small, and much smaller in practice than the four shillings expected as royalties.

"It must not be forgotten that the royalties are the first duties on product, and that they precede the profits, and that Iraq shall receive royalties from crude oil whether profits were effected or not; in fact, Iraq shall definitely receive royalties a long time before any profits are effected... This means that Iraq shall secure from the beginning a steady income instead of an indefinite hope of future profits. *And the British Government feels that owing to the present financial situation of Iraq, it is preferable to secure a steady income rather than to have an indefinite hope of profits.*

"Therefore, we find that, for the benefit of Iraq (sic), the Government should accept the royalties of four shillings and forgo the demand to share in a part of the capital. *I believe that the Iraqi Government should accept the advice of the British Government* on this point."

It is sufficient to quote that much from the confidential letter sent on 24 December 1924. Five days later, H.E. Muzahim Al-Pachachi sent a reply refuting the contents of that letter to the Prime Minister. On 20 January 1925, a negotiation session was held, chaired by the British High Commissioner. He declared that, "It is quite logical for the companies to claim that there is a connection between the question of sharing in the capital and the question of royalties (that is to say that participation should inevitably lead to diminishing royalties for the Iraqi Government) but the Minister of Finance and the Minister of Communications and Works were not convinced of this argument".

The High Commissioner added that he was "certain that there was no possibility whatsoever that the company would consent at the present stage to the Iraqi Government exercising her right to share in the capital so long as French interests, or probably American interests [British interests are naturally innocent of this obstinacy! ] found that this would lead to disturbing the balance existing among different interests concerned. Therefore, *if the Iraqi Government insists on exercising this right, it will be necessary to declare the end of negotiations".*

What was the reply of the Iraqi side? What was the Iraqi comment on this final stage of negotiations?

**Very Urgent**

No. 93
Ministry of Communications and Works.
Baghdad, 16 February 1925.
To the Secretary, The Council of Ministers,

Dear Sir,

With reference to the decision taken by the Council in the session of 14 August 1924, and to your letter No. 921-4-7 dated 16 August 1924.

The Ministerial Committee formed of the undersigned has looked into the suggestion made by the Turkish Oil Company, and has negotiated with the company representative in Baghdad, Mr Kittling, in various sessions held for this reason... no agreement could be reached on the following points, as the opinion of the Committee thereon was not accepted by the company representative. These points are of two types, those of major and those of secondary importance.

The first type are:

1. Concession rights in 24 pieces mentioned in article NO. 5 of the draft agreement.
2. Confirmation of government rights or the rights of Iraqi concerns in sharing in the company capital up to 20% as per article No. 8 of the San Remo Agreement concerning oil.
3. Registration of by-companies mentioned in article No.33 of the draft agreement in Iraq as Iraqi companies.

Concerning the first point, the company agrees that the government take her place in article no. 6 [i.e. the government takes the place of the company outside the 24 pieces chosen by the company], but specifies at the same time that the company should act as agent for the government in all matters (concerning these areas) and that all sales should go to the company and not to the Government in any way [i.e. the state's right to act became a mere formality].

Concerning the second point, it has been agreed that the right to share in the company capital up to 20% is given to Iraq, as per the San Remo Agreement, which the British Government considered among international treaties and agreements referred to in article No. 10 of the British-Iraqi Treaty. And, with reference to the letter of H.E. the High Commissioner No. B-O-64 dated 24 March 1924, addressed to H.E. the Prime Minister, it becomes clear that the company mentioned in article No. 8 of the San Remo Agreement is meant to be the Turkish Oil Company. This means that the said company had undertaken to abide by the article of this agreement on equal terms, without preference of one article over another. Therefore, we do not see any reason for the company to refuse the international acceptance of this right. At the same time, we see that the British Government finds a strong connection between the government share, specified in article No. 10 of the concession agreement, and the right of the Iraqi Government to share in the company capital, and finds that the Committee's insistence on this right calls for a decrease in the government share meaning the four shillings – which action affects the revenue of the Iraqi government. On the other hand, the British Government further sees that convincing the interested parties in the company of this is not possible at present, and believes that insisting on this point would lead to a cessation of negotiation.

The Committee does not find the least connection between the government share (the royalty) and the right to share in the company, as was specified by the terms of the San Remo Agreement. The government right to royalty comes from offering the company a concession to develop oil in Baghdad and Mosul *wilayets* [provinces] and there is no difference here between royalty and fees or taxes levied by the government from farmers and landowners for their use of water or for land development, like the one-tenth and the one-fifth shares, etc. We do not believe that oil is less important or less significant than water or land for the Government not to have a similar share from its development. As for the monies which the government may be owed from company profits, this will be in return for capital and effort put into the company and nothing else. What confirms the fact that the royalty is not participation in capital is article No. 8 of the San Remo Agreement. The right to participate belongs to the government or the citizens. The Agreement's use of "citizens" clearly shows that there is no connection between the government right (the royalty) and the right to share in the capital. Therefore, we find that the connection is completely non-existent between the two rights, and the reason behind the one is clearly different from the other.

Concerning the third point, the Committee fears that after the British-Iraqi Treaty comes to an end, and the ancient concessions are reinforced, great problems may arise with the by-companies working in Iraq when they demand enjoyment of these concessions, which may affect both Iraqi Government and citizens. To avoid such an occurrence we find it necessary to stipulate that such by-companies be registered in Iraq, so that they become Iraqi companies from the start, since the promises of the British Government to abolish these concessions are dependent upon the application of terms most of which are ambiguous and indefinite, and we do not think that such terms are binding to Britain in any way, or that they are completely applicable.

We hope this report could be submitted to the august Council at their earliest convenience to reach a decision on the concession as the Council sees fit.

Respectfully,
Minister of Communications and Works
Minister of Justice
Minister of Finance

This wonderful disputation outwitted the adversary by showing a clear understanding of the structure of relations that must exist between the sovereign state and the oil monopolies. The Iraqi side at the negotiations had defined the basic points that continued to be the central issues which patriotic forces continued to fight to gain from companies: limiting concession areas, state participation in ownership and administration of companies, and companies' subjection to the sovereignty of the Iraqi state.

In other countries, several decades later, there were those who signed the agreements without even reading them. There may be some justification for that, like ignorance or extreme backwardness. But whatever the reasons, this firm and conscious attitude in the Iraq negotiations, fifty years ago, denotes the depth of the national movement, and nominates this country for a prominent part in the struggle of oil-exporting countries.

Yet, concerning the negotiations and communications of 1924-5, we should not forget, in the long run, that picture we presented at the start about the status and personalities of the negotiators. Negotiations in these matters do not form an intellectual forum where victory is for the right, or the more sound, argument. Negotiations in these matters are an expression of a struggle between powers, and the result of negotiations cannot be but a reflection of the strength of these powers in the struggle.

On 7 February 1925 H.E. Sir Henry Dobbs, the British High Commissioner, sent a confidential letter to the Prime Minister, explaining his official point of view about the Turkish Oil Company (TOC) concession, to be submitted to H.M. The King and the Council of Ministers. Part of the letter read:[4]

> "Now I come to the political situation, to which Your Excellency alluded. The Frontier Committee has shown by its attitude and questions that it attaches great concern to the Iraqi decision about the concession of the Turkish Oil Company. That attitude finds that if there is a great number of international groups with an interest in the oil in Mosul [Mowsil] and Baghdad *wilayets,* then this interest in itself implies active security to support the stability of the Iraqi state and to defend her against aggression."

H.E. the High Commissioner frankly connects submission to TOC to the decision of the Frontiers Committee of the League of Nations about continuation or separation of Mowsil as part of Iraq. He also implies that refusal to sign the concession contract would anger large international groups which would make Iraq open to aggression. As usual, it was necessary to stray from the subject, so the High Commissioner digressed:

> "I am not concerned here with a justification of the Frontiers Committee in this connection. I am only interested in attracting Your Excellency's attention to the facts. It seems vitally important that the Iraqi Government should give a good impression to the Committee at an early stage of the Committee's investigation, by proving that these international groups (connected with the oil company) are concerned with the question of stability.
>
> "Finally, I hope Your Excellency will take into consideration the attitude of the British Government in this matter, which, after long inspection, has found that the terms suggested by TOC are more agreeable to the Iraqi Government than any terms that could be obtained."

This is an open threat that Britain, with all her power, was with the company, and would of course stand against whoever opposed her will.

In fact, when the Investigation Committee arrived in Mosul on 27 January 1925 and started to question the citizens about the future of that *wilaya*, it became clear to the Iraqi government that the League of Nations would really not allow this *wilaya* to remain within Iraq unless Iraq offered TOC a concession to search for oil reservoirs in this debatable *wilaya.* Under this and other pressures, the Council of Ministers held a meeting on 26 February and issued this decision:

"The Iraqi Government, while not recognizing that TOC has any concession, is acquainted with the promise mentioned in the letter of the Turkish Prime Minister to the British Ambassador, dated 28 February 1914. The Government is ready to fulfil this promise, provided the company agree to the terms prepared by the Iraqi Government. The Iraqi Government is not willing to postpone the decision about the concession and there is no basic point pending except the question of shares (i.e. participation in capital) which is now under discussion."

The statement was not sufficient, and the British Government declared her refusal to ratify the Iraqi Constitution which was previously ratified by the Iraqi Constitutional Council, so the Council of Ministers retreated on 5 March 1925 and stated its approval of the broad lines of the British-Iraqi Petroleum Agreement, and dropped its reservation about the question of participation.

The Minister of Works and Communications had to swallow his previous opposition, and was asked to sign the Agreement for the Government, and he did.

True there was strong opposition, even inside the Council, and two resignations: that of Mohammad Rida Al-Shabibi, Minister of Education, and Rasheed Ali Al-Gailani, Minister of Justice. But the Agreement was signed on 14 March 1925, and became valid.[5]

**NOTES**

1. The San Remo Agreement had organized relations between Britain and France in developing oil in Iraq, providing for the right of Iraqis to participate in the company's shares to the value of not less than 20% of total shares. This agreement was considered among international agreements binding to Iraq, and consequently to all three parties: Britain, France and Iraq. But the companies later refused this principle, and, when drafting the agreement signed in 1925, mentioned the principle (M-34) where implementation was made dependent upon issuing shares to the markets. At the same time, the companies organized themselves as non-share companies, that is companies which do not issue shares, thus taking away the right of Iraq to participate. This subject remained a point of dispute.

2. *Oil Documents in Iraq* Vol.1, translated and edited by Q.A. Al-Abbas, (Baghdad, n.d.) pp.303 and 6. (Arabic).
3. *op. cit.*, pp.326 and 329.
4. *ibid.*, pp.321 and 323.
5. M. Al-Mousawi, *Iraqi Oil,* (Baghdad, 1973); also O'Conner, I. Allawi, *op. cit.*

## CHAPTER 7

# OIL AND THE HIGH STRATEGY OF THE STATE

On 15 January 1972, the picture presented by the city of Baghdad was completely different from that of the Twenties. It was much larger, and much more beautiful.

In the evening of that day a round of negotiations began between the representatives of the state of Iraq and the representatives of IPC. Every detail of the picture of the negotiators around the negotiation table tells us that everything had changed. The Pashas and Begs are no longer in the picture. The place was a meeting-hall which was elegant, with touches of arabesque – the National Assembly overlooking the Tigris.

The only thing which had not changed with time was the mentality of Mr. Stockwell. He was carbon copy of Mr Kittling.

In the time between the last chapter which occured in 1920's and this final round of 1970's world wars had broken out which had liquidated empires, brought great states down, and carried others to the top; a group of Socialist countries had been formed, radical revolutions had succeeded in a number of colonies and semi-colonies of the past; the rest had achieved political independence... In our Arab area the revolutionary ebb had run high, and old bases and pacts were liquidated. Nationalizations of reactionary and imperialistic interests came one after the other, beginning with the Suez Canal, and expanded, aiming at reactionary government systems some painful blows and even sometimes fatal ones.

Amidst all this the Iraqi Revolution broke out in 1958 and the new rulers faced up to the oil monopolies and demanded the right to negotiate to modify the structure of relations between monopolies and the state of Iraq, and were faced with refusal, exactly as the negotiation delegation was faced with refusal in 1924. When Iraq resorted to unilateral legislation by issuing Law No. 80 to emphasize sovereignty over Iraqi land and systems, the companies insisted on refusal, and stressed that they were an entity out of the reach of local state powers.

True, the differences among patriotic forces and spasmodic convulsions in the revolutionary process experienced by Iraq since 1958 had encouraged the companies in their manoeuvering and arrogance, but it was necessary for the companies to realize in 1972 that the passive aspects of the political situations, which the companies had exploited since 1958, were no longer in control of the situation.

This round of negotiations started three and a half years after the 17-30 July Revolution. During those years, the ABSP proved that it had comprehended the lessons and experiences of the bloody period between 14 July 1958 and 17-30 July 1968.[1]

The ABSP had seized power on July 17, through an operation planned by the regional command of the Party.

> "The plan of the Command was to include the direct participation of the members in attacking the Republican Guard and bringing down the system, in addition to the participation of a number of retired military comrades, plus a number of civil comrades. This stemmed from realizing the necessity to emphasize a question of great importance in the life of the Party and in the process of the revolution and its future course. The Party Command must not be a command in thought, direction and planning only. It should also be the vanguard of the practical process, bearing all hazards, which will put the Command in direct contact with reality, with all its needs and developments. Any future decisions would thus be realistic on the one hand, and, on the other, the Command would gain full esteem which should characterize the leadership of the Revolution. This would also stop, from the beginning, the possibility of having two extremes in the revolutionary system: the extreme of the intellectual and political leadership, on one hand, and that of the men of action, who deal with the actual events and their practical developments and hazards, and alone claim the honour of attack, on the other."

This lesson is of prime importance and must be comprehended by all revolutionary powers in the Third World.

After seizing power all people who were thrust upon the new command who were at all questionable and could have turned July 17 into a mere *coup d'état,* were removed on 30 July with precision. So, the power was totally in the hands of the Ba'th Party, represented in the Revolutionary Command Council (RCC) and the Council of Ministers. But establishing revolutionary power needed something further and more difficult to attain.

> "The course of establishing revolutionary power, and securing its real leadership by the party, was a difficult and highly complicated job. It demanded following a balanced and gradual plan with various versions, which, though apparently different at times, were

linked with one thread which was the future image of this power."[3]

"The Ba'th party had to face up to difficult jobs in these fields. The balance of power in the society on the one hand, and the subjective and objective situations concerning the Party, the country and the world on the other, made it necessary to employ the system of fast and decisive attack on power at the top, in order to bring it down and control it within hours. So the Party found itself, in a few hours, in power. The revolution had not had time to build its own system before coming to power, as was the case with some revolutions in the world.

"The balance of social powers, which necessitated these methods in order to seize political power, the nature of circumstances in Iraq and the Arab area, the circumstances of the modern world with its rapid and advanced systems of communications, war, production, services, publicity and other vital fields make the state an indispensible tool even for one day, a tool which cannot be paralysed on a large scale. Therefore, to dismantle that system in its place, or undertaking immediate changes on a large scale will undoubtedly lead to comprehensive chaos and huge damage.

"In spite of these dangers, the Party and the Revolution had to maintain the apparatus and concentrate at first on tightening control of the centre of government and key positions in it. It was by a long, complex and gradual process that the necessary changes were introduced throughout the state apparatus, as well as in legislation, the information media, culture and education. In this way, conditions were established for radical change in accordance with the Pan-Arab, democratic and socialist principles of the Revolution."[4]

This can be seen as another lesson demonstrated by the Iraqi experiment in the form of an important theoretical contribution about the ways of controlling the state system. In its general direction, this does not represent a mere local experience, but it is applicable also in numerous similar cases.

In accordance with this sound theoretical understanding of the circumstances, handling the army in particular becomes an exceedingly difficult job. The revolutionary seizure of power, with the participation of military personnel, may leave the door open for divergent political practices in the armed forces generally. Among such military personnel some adventurers may emerge to monopolize the Revolution by force of arms. The revolutionary uprising in itself may attract others, who did not take part, to try and repeat the seizure in a mere *coup d'état*.

The Ba'th Party Command faced that challenge, armed with previous bitter experience.

> "The Party and Revolution faced from the very first day, and firmly, two basic questions [à propos the armed forces].
>
> "The first: to strengthen the Party Command of the army in addition to purging it from questionable, conspiring and adventurous personnel; to propagate the Party principles and the general national and socialist culture among its members; to lay the bases of military and organizational limitations which enable the army to fulfil its duties in the best way, and to safeguard against slips and diversions committed in its name by the military aristocracy of the Qassim and Arif eras; to secure complete alliance with the popular movement led by the Party, sound and actual participation in revolutionary construction, and in undertaking patriotic and national duties.
>
> "The second: to terminate the situations of chaos and backwardness which rose in the army during past eras, and to re-organize it along modern principles and to develop methods of training, recruitment, and technical and fighting capacities, to increase its size and provide it with strong and modern weapons and equipment."

To carry out these two jobs was difficult. The first job in particular demanded, in accordance with the previous explanation of general principles, a long and complicated process.

> "Therefore, the talk about disbanding the old army and building a new strong army, or hastening to undertake large scale changes in the army is far removed from the objective evaluations, and does not express a serious revolutionary attitude, as it implies a great avoidance of the needs to achieve the sought objective."

Why? There is a general reason to be added to those reasons connected with considerations of a more detailed or technical nature.

> "The nature of modern armies, their weapons and techniques, is widely different from that of the armies in the early decades of this century. In the past, the main weapon of those armies was the rifle, which did not need more than a few days of practice. But the weapons of modern armies are varied, such as planes, armoured

vehicles, artillery, radar and other equipment of a highly technical and complicated nature. They consequently demand many years before soldiers and officers can perfect their use.

"In addition to this, we must take into consideration another basic and important fact, and that is the comparatively small geographic area of Iraq. This is of exceptional importance in this age, which is characterized with highly developed far-range weapons, and exceedingly fast communications which form a great danger to a country's security if its army remains weak for a long period of time. Therefore, adopting the long term method of rebuilding the army on a strong revolutionary base was the only practical way available to the Party."[5]

But, it is, of course, a method which calls for constant vigilance and a hand on the trigger that knows no hesitation in the time of need. This is the aspect which can be applied to other similar situations. In the case of Iraq in 1968, the Party was practically faced with a state of quasi-civil war because of the situation in the northern area of the country. A part of the Iraqi army was camping in the eastern front in Jordan – more than 5000 troops – and before the Revolution completed its first year of life, the Iranian government abolished the 1937 Treaty and there developed a serious military threat on the eastern frontiers.

The importance of all this does not only lie in the fact that it is a version of a theoretical principle which seems to be thus proven. More important than this, which emphasizes the previous statements, is that it was practised with ability and it achieved the aims. The leadership of the political establishment was upheld in all the departments of the state, including the military, thus protecting the strength of the central revolutionary power against any attempt at a breakthrough.

But the establishment of revolutionary power, in spite of all that, cannot be limited to operations inside the state system, and separate from social struggles and class upheaval and change of political power in various levels and directions. It was known that Iraq was,

"thickly stacked with espionage networks: American, British, Israeli, etc... Those networks had penetrated into the armed forces, security systems, economic establishments, some political and religious movements and other sensitive centres in the state and society which had led to Iraq becoming almost an open field for imperialists, Zionists and reactionaries... The activity of those networks was not limited to collecting information on political, military and economic matters and passing it over to the enemy; the main objective was the immediate seizure of power, or indirectly influencing that end by supporters of imperialism and reaction."[6]

In this field, the Revolution resorted to decisive action against enemies, which needed to be violent, even exceedingly violent. But, what about the other side of the coin? What about the relationships with forces that were supposed to be allies?

Here we come to the greatest achievements of the 17-30 July Revolution in establishing the foundations of progress. The Revolution had fully comprehended the central reason behind the previous relapses of the regional and national struggles.

"There were serious basic reasons, at the forefront of which stood the deviation of Arab progressive forces in allowing internal contradictions to take the upper hand over the basic contradiction between those forces as a whole on the one hand, and imperialism, Zionism and reaction on the other, and to accelerate those subsidiary contradictions to the level of serious conflict, even war."[7]

Starting from this idea, the Ba'th Party, after 30 July, called for the necessity of alliance among patriotic forces, only a few months after coming to power, when Tariq Aziz published a wonderful article in *Al-Thawra* in December 1968 expressing the Party attitude, under the title "Towards a Firm Progressive Patriotic Front". These are some excerpts from that article:

"If the national action prior to the 14 July 1958 Revolution was characterized by joint struggle under the banner of the national union front against the reactionary royal regime and its connection with imperialism, the general progress of the national action after that revolution, and for ten years, through shortlived and unsuccessful experiences, was characterized by the differences between each patriotic force in action, with violent struggle among patriotic forces, and even with division and warring among the ranks of each of those forces.

"What was the result for the patriotic groups and the people in Iraq and the Arab world of that course of action?

"Each of the patriotic groups came to power with many brilliant mottoes and unlimited aspirations, but did not remain long in power, leaving it amidst the malevolence of all other groups, the doubts of the people, and a huge harvest of errors, until power at last came, before the 17 July 1968 Revolution, into the hands of a group that had no connection with any of the other patriotic forces, or any roots in past patriotic movements of any type... a colourless group with no real identity, afraid of the patriotic forces and a suffering deep alienation from such forces and from the people, coveting only power and managing only the affairs of the day, looting whatever could be looted of the riches of the country which became transformed into a farm for relatives, friends and supporters. Finally the mere existence of that group in power, irrespective of the different evaluations and analyses of it, became in itself an insult to all patriotic forces and to the entire people of Iraq. ... Until the 17 July 1968 Revolution, no form of real union among patriotic forces had occurred. On that date, and particularly on 30 July, the ABSP put an end to the former era and attained power. From the position of power, the Party called for co-operation among progressive patriotic forces.... Some may imagine, that talking about the Front today is no longer relevant, and that it might have been so two or three months ago, but that appearances nowadays show that each of the patriotic forces has returned once more to take a position behind high stone walls opposed to the other groups. In fact the talk about the Front is relevant every day, even every hour. The question of the Front is not seasonal, nor is it a question which we may discuss today and forget tomorrow. All patriotic groups consider the Front as the corner-stone of the strategy of this stage in the revolutionary process.... In order to form the real Front and define the Pact, it is necessary for the Revolution and the patriotic forces not in power to follow the means that will pave the way to the final goal. As for the revolutionary system which called for a union of the patriotic forces, that system must always show by word and practice, that this experiment is not a mere outcome of romantic emotions, good intentions and the avoidance of monopolizing the power. It is in fact a call stemming from a deep-rooted belief that the objectives of the progressive national Revolution, with a socialist and unitarian horizon, especially in the circumstances of confrontation with the Zionist imperialistic aggression, are more than can be achieved by one group alone....

"As for the other patriotic forces, it is a mistake on their part to look at the existing power as the power of the 'other party' that must be weakened and rendered of little value. This, naturally, does not deny the right to criticise or mean that they should give up attempts to develop. But what the patriotic forces should not forget is that weakening the government will not be as advantageous to those forces as it would be to the forces of reaction.... When the potential and the experience of all progressive and patriotic groups are put together, they will find enough difficulties in achieving revolutionary aims of democracy, liberation, and progress. What would the situation be like if we asked only one patriotic group to try and achieve everything? The sound approach is for all progressive patriotic forces to unite in achieving progressive and democratic aims, even before the final agreements are made on Front and Pact. This plan of attack is feasible in many cases."

In the light of these concepts, and in the name of the ABSP, Tariq Aziz advanced four such joint objectives:

1. Defending the existing order in every way against imperialistic conspiracies and the stratagems of reactionary forces.

2. Acting directly and indirectly, and with whatever means at the disposal of the patriotic forces, to solve the Kurdish problem, since this was a basic obstacle in the way of forming a real patriotic front.

3. Participation of men and ideas in the independent oil policy of the Revolutionary Government. He said that any talk about a progressive national Revolution, with the oil remaining in the hands of monopolies, and Iraq remaining unable to exploit its oil and mineral resources in a direct manner bringing large returns to enable Iraq to be liberated from the threats and manoeuvres of foreign monopolies – such talk was mere nonsense.

4. Participation in every possible way in the process of harnessing the military, political and economic powers of Iraq for the imminent battle with the Israeli enemy and its imperialistic supporters.

This clear vision about the importance and objectives of a national patriotic front was not at all easy to carry out in practice, and the Ba'th Party had no illusions about that. The Eighth Congress Report says that

"The ABSP has faced, like other parties have faced, heavy difficulties to provide a positive psychological atmosphere for the Front's actions. The problem was not theoretical; all progressive political parties were theoretically convinced of the need for action within a front. It was rather a psychological and behavioural problem, as the actual situation in Iraq always caused problems for many reasons, two of which are the general popular temperament, and the strong influence of former warring relations. Experience shows that although it is true to say that the psychological and behavioural aspects should be subservient to the theoretical analysis in the work of the political parties, reaching that result was not an easy thing, and it demanded a long time and exceptional efforts."

In fact, it was not easy at all to build up a strong patriotic order. Building such an order had to rely in turn on the unity, or at least on a mutual understanding, of modern progressive forces, if what was wanted was to achieve radical changes in society, a termination of economic control of imperialism, and a liberation of the society from backwardness. Backwardness cannot be abolished by a few laws being passed about nationalization or agrarian reform. These progressive measures alone are not enough to liquidate tribal connections, bigotries and contradictions which had gone deep into the conscience of various sectors of the population.

Undoubtedly, faced with this backward heritage, and alert to outside influences, and the need to prepare for independent development, political and progressive forces need to instigate a cultural and publicity revolution. This should endeavour, through a programme of comprehensive change, to replace the traditional differences in society by the basic political differences which exist, in fact, between totally patriotic forces on the one hand, and imperialistic, reactionary and backward forces on the other. This is the duty of all modern progressive vanguards, which, through their collective efforts, can contain and overstep the traditional divisions in order to undertake a decisive struggle in the face of the real modern conflict, and for the sake of an integrated revolutionary jump that will place the society in a new orbit of development.

There is nothing new in this. All comrades of progressive modern vanguards realized its importance. The experience of bitter in-fighting among those vanguards throughout the ten years of national and patriotic failure in Iraq from 1958 to 1968 had ripened and crystallized the alliance issue. But the same bitter experience also meant that there were feuds among the parties. And whatever may be said about the maturity necessary in the vanguards of the new society, they are – in the final calculation – human beings too.

This does not mean that understanding leading to unity is an impossible step, but it means that the effort demanded is huge. If the leadership does not despair from making this effort, and does not lose its direction under pressure, then this leadership is genuine and deeply revolutionary. Since the insistence on achieving the Patriotic Front was indicative of this, what made the action easy was the gradual development of mutual confidence. Again we refer to Tariq Aziz's article:

> "The formation of a firm patriotic front in Iraq, and the complete agreement in a serious pact for national action in a short period of time are two matters which – to my mind – reach the level of a miracle. Even if a front and a pact were achieved in a short period, they would not be standing on firm ground, and would always be liable to various kinds of hazards.
>
> "Sincere wishes cannot create a real patriotic front, even if objective circumstances denote the necessity of its creation. In addition to objective circumstances and wishes there must be two conditions: first, analyses and sound definitions of the structure, objectives and method of action in the Front; second, actual practice of thinking and acting within the Front on all levels, and for a sufficient period of time, so that this practice becomes a permanent general quality of patriotic action throughout a certain historical stage....
>
> "Therefore, the practical and safe way to reach to the Front is to follow in the steps of national co-operation, gradually gaining in faith, until we reach the level of setting up the Front and defining the Pact."

These clear concepts were not mere sweet words uttered to clear the conscience, then forgotten. These concepts had an iron will behind them. The Ba'th Party presented these concepts when at the peak of power, and the progressive forces responded. Amidst numerous difficulties, progress was slow, and on the way a rugged and concentrated discussion began on the subject of political leadership. On 11 March 1970 the Ba'th Party Command announced a statement recognizing the national rights of the Kurds and providing the principle of autonomous rule in the Kurdish area.

> "The Party's success in presenting a "theoretical" and "political" programme at the same time confirms the genuine democratic aspect of the national theory of the Party on the one hand, and, on the other, confirms the practical ability of the Party to solve a question of this nature, despite its critical and complicated aspects.
>
> "Also, the Party declaration of March 11 spelling out the 'theoretical' and 'political' solution of the Kurdish question, was

considered a unifying factor in the national movement in the country, since that question used to be a basic cause of dispute inside the movement. So one of the greatest impediments in the way of setting up the porgressive front in Iraq was removed."[8]

The statement was not sufficient to solve the problem. That problem could not be radically solved except through a comprehensive economic, social, political and cultural revolution. Such a problem could not be solved by issuing statements or legislations. But the value of the statement is that it showed the way and laid the foundation.

Has all this taken us away from the subject of oil? Never. We are on the very subject, and on the subject of nationalization in particular.

In a game of chess, you cannot say about the last move, the checkmate, that it was the only move which won the game. All the pieces participated with co-ordinated moves, and in accordance with calculations that did not lose sight of the adversary's calculations. They all took part in the checkmate.

The same thing is true about any war, hot or cold. Neither of the fighting parties can say that the direct confrontation on the battle field was the only factor in victory or defeat. Separate military battles do not decide the encounter, however good they may be, if the military strategy behind these operations is not up to satandard. Also, the military strategy will certainly fail if the higher strategy of the state does not help. The success of military strategy demands action on an international level to upset the enemy alliances and support friendly ones. The success of military strategy also needs to secure provisions, arms and munitions. Economic fortitude is another way to face the attrition of war and continued enemy attempts to ruin sources of production. Before all this, the maintenance of the people's morale at a high level, the readiness of the masses to give support for the cause of the war and the fighting forces, is the first and most important condition. Forging a well-drawn military strategy is essential; trained and equipped soldiers, commanders efficient in directing the operations... all these are matters of unquestionable importance. But alone, they do not decide, or rather they are not the most decisive. Therefore, the achievement of victory in war is not a matter for military command with all the genius and greatness of the role undertaken. Victory is dependent on the high command that runs the entire encounter in the first place.

This high command is expected to have a comprehensive view, extending from the international and regional politics, to the local political situations, to following the economic developments, and following at the same time the developments in the military situation. And this high command is expected after this to use its comprehensive view in all fields, with dexterity and courage to further the aim: defeat of the enemy.

This applies to every war, and a "hot" war is not different here from a cold war except in priorities. But, undoubtedly, a war between two equal powers is different from one between a large wealthy power and another power which seems materially less strong. The latter is exemplified by the Vietnamese liberation war against the imperialistic dinosaur. The difference in such a war will be in the organization of the factors which give strength to each side. Those factors differ, and each side has to form the ideal fighting machine from the resources available. But however you juggle these resources, how can a small nation, economically and technologically backward, how can such a nation in any way form an ideal fighting machine equal to and able to face up to what the US can use?

This question was easily answered for the imperialistic American mentality, rather for all those who use their minds like a computer. The US owns types and quantities of arms many times greater in firing power to those that could ever be found by the Vietnamese. True that this war would be a burden on the American economy (though the expectations about the size of that burden did not forsee what the burden would actually prove to be). Yet, the American economy, in any case, could bear the war expenses. But what could the Vietnamese do when all their resources and producing establishments were ruined?

No outside aid would ever fill the gap between the American and the Vietnamese material powers. No aid could change the fact that the economic exhaustion of the Vietnamese would immeasurably exceed any troubles faced by the US economy. Whoever calculated according to these data, which appeared quite sound, came out with a conclusion which he thought decisive. Whoever considered himself "sensible" made light of the Vietnamese foolhardiness. Whoever thought himself a friend asked the Vietnamese leadership to avoid foolhardiness and strong-headedness.

Now that we know the outcome of that struggle, it becomes imperative that those who estimated wrongly should analyse the causes of their mistakes.

The struggle of a small nation against the oil monopolies, against the "terrible" international cartel, is a kind of cold war that can turn "hot". In this struggle, the previous rules apply. The question is not in the hands of the soldiers on the battlefield. The possibility of victory or defeat will not be hanging round the necks of the technicians who run the production operations, nor round the necks of those who study and follow the oil economics in such matters as production cost and price, or marketing economies. The role of all these people is important. But any strategy for a national oil struggle must be subject to high strategy of the state as a condition of victory. The efficiency of the oil strategy and the dexterity of those who put it in practice do not suffice if such strategy is not directed by a comprehensive political strategy, led by a dexterous and courageous command. Therefore, we cannot regard the success of direct national development or the later nationalization of operations in Kirkuk as something separate from what had happened in the state structure, and in the relations among patriotic forces or in the field of international relations.

In the war of economic liberation the same criteria apply as in the case of the armed liberation war of Vietnam. How did Vietnam win? How did it balance its own capacity so as to equal, then exceed the material capacities of American imperialism. The Vietnamese experiment proved that moral power was a powerful energy that could replace material deficiency, thus shifting the balance.

This energy cannot be gauged by an outsider, but the one who inspires that energy can gauge it. The moral energy can go infinitely wider and deeper, and the revolutionary command alone can realize those dimensions.

The Iraqi leadership was able to realize the moral energy of the people. Every Iraqi had been living with a personal desire to settle the accounts with oil monopolies. Everybody knew that those monoplies were looting, and that they were behind all the catastrophes that checked progress. On 31 December 1971 Saddam Hussein said: "When we can live on half salary for one or two years, then we can make oil companies bow", When he said that, everybody was waiting for the signal, provided the man giving the signal was trustworthy, and provided the non-essential problems which plagued unity of will be solved. The preparations and achievements of the years following the 1968 revolution undertook to realize these conditions.

A round of negotiations then began with the oil companies on 15 January 1972. The Iraqi delegation was sitting round the table, backed with a revolutionary political leadership, which was in turn backed by strong national power, which could put an end to agents of imperialism and reactionism, and could take large steps to form a front joining progressive patriotic forces. The Iraqi delegation represented a leadership which had liberated its international political relations from pressures and the directions of imperialist states, and had strengthened its co-operation with Socialist countries, and confirmed its attitude as a non-aligned power. Between the home political achievements and what was achieved in foreign fields, the insistence on liberating oil wealth was apparent everywhere, and the decisive achievements in direct development did not falter.

In 1966 an IPC representative had said spitefully: "A great deal of blood was shed in Iraq, but, by the grace of God, not one drop of oil was shed". In 1972 the patriotic forces had completely stopped the bloodshed, and the wounds were almost healed. How did the IPC not realize that it was her turn to lose blood?

## NOTES

1. Conversation with officials.
2. Political Report of the Eighth Regional Congress Part II.
3. *ibid.*, Part III.
4. The 1968 Revolution in Iraq. The Political Report of the Eighth Congress of the Arab Ba'th Socialist Party in Iraq, 1974. (Ithacca Press – London 1979) p. 110.
5. *ibid.*, Part VI.
6. *ibid.*, Part IV.
7. *ibid.*
8. *ibid.*

# CHAPTER 8

## THE LAST ROUND

"27 November 1971
Ministry of Oil and Minerals
Ministerial Bureau, M.H.S. 1640 *Strictly Confidential*
Iraq Petroleum Co. Ltd., Basra Petroleum Co. Ltd.,
Mosul Petroleum Co. Ltd, The General Representative.

"Gentlemen,

"With reference to the interview between you and the Director General of Oil Affairs, on 20 October 1971, about the opinion of the companies on the bases of negotiations with the Government, which may be summarized as follows:

"1. The companies undertake not to ask the Government to cancel, annul or amend Law No. 80 of 1961, openly or implicitly by legislation.

"2. The companies are free to study the results of the said Law on their operations in Iraq...."

We notice that paragraph two may practically cancel out paragraph one. Equivocation is an art practised by the companies. We should not forget that the companies used to state – even in the 1924 negotiations – that they did not deny the legal right of Iraq to own 20% of the shares, only to withdraw that approval in the following line in one way or another. In any case, and to avoid the customary equivocation, Dr Sadoon Hammadi, the Minister of Oil and Minerals at the time, had to digress in his letter and answer that possibility.

"1. The Iraqi Government appreciates the prior undertaking of the companies not to request the Government to issue any legislation cancelling or annulling law No. 80, openly or implicitly.

"2. The Iraqi Government sees that the important question is not concerned with annulling or amending the Law, since that is a right exclusively limited to the sovereign state. No government can accept any request about cancelling or amending her legislation, a matter normally taken for granted but which does not seem to be so with the companies. What the Government meant from the start does not concern the legislative question, but is concerned with the protection of the content of Law No. 80.

"This is the important point. As for studying the result of the Law on the companies, the Government sees nothing against such study, provided it does not lead to suggestions and arrangements which may weaken that content; that is to say, the Government, under any circumstances or in any case, cannot accept any arrangement to settle the pending questions which may rob the said law of its content.

Respectfully.
Minister of Oil and Minerals

A clear and definite preparation on the Iraqi side, in accordance with instructions from the Follow-up Committee. At the same time it was decided to re-form the Iraqi delegation to the negotiations. The earlier one was technical, headed by the Minister of Oil. The new delegation was meant to be political.

Then the first meeting began, after some communications had been exchanged in the evening of 15 January 1972, and it was a heated beginning.[1] The chief of the Iraqi delegation started the session by saying:

"The degree of objectivity and new understanding of our situation and methods which you show during the negotiations will make the job easy and will be appreciated by all. It will also be in the interest of your companies and will save a lot of trouble for the parties concerned... You have been following a development and production policy based on continuous pressure on Iraq in order to weaken her, and have obstinately refused to pay the royalty expenses due, which are a clear and recognized right for all oil producing countries, and have constantly interfered in production averages in our oil fields.... I do not so much want to blame the companies for their attitudes which led our relationship to such a high degree of tension and lack of confidence. But I ask you to respond to our rights in a realistic spirit, conscious of the new situation set up by the 17 July 1968 Revolution under the leadership of the ABSP.... The rise of a revolutionary rule depending on a solid popular base is an important question, which we wish you could consider and we will refuse any bargaining logic and any method which aims to harm our political and economic interests."

After this general survey, the accent of which reflected that the Iraqi side was conscious of the drastic change in the balance of power, the head of the Iraqi delegation surveyed the Iraqi claims: some basic, others secondary.

What was the reply of the leader of the negotiation team of the companies, Mr. Stockwell? He completely ignored the political references and the strong words used by the head of the Iraqi delegation, like pressure, insistence, interference, tension, lack of confidence. Very coolly, he said: "I am in complete agreement that we should be serious in deciding the issues". Then he chose a subsidiary issue in the Iraqi claims and began his discussion.

These early minutes of the negotiations are indicative of what went on after that in the sessions. The Iraqi delegation insisted that the round should be decisive, and saw – for the first time – that the desired changes in the structure of relations with the companies had to be enforced, and that Iraq was not only able but was prepared to enforce them by unilateral legislation if negotiations failed. In accordance with this decisiveness, the Iraqi delegation led the course of negotiations from beginning to end. The companies, on the other side of the table, were shrugging their shoulders to the Iraqi delegation from the early moments, imagining that negotiations would not veer from the customary framework. The "host" country should not step outside the role of someone who was getting some money, and within this role there should be no objections to making some concessions in one way or another.

Yet, what provoked the Iraqi side is that even in the traditional field of conflict, the field of the rights of the state as a money-collector, the companies tried to devour that money or bargain audaciously about recognized rights. The issue picked up by Stockwell to start negotiations was the question of expending royalties. This was a valid principle in all OPEC countries except Iraq. Irrespective of its validity in other countries, royalties are not a gift from the companies to the producing countries, and it was not a right which had emerged merely because it was decided by OPEC. In the words of Oil Minister Dr Sadoon Hammadi in the second session: "We have a scientific and realistic basis for royalty payments and that is that the payment thereof to the land-owner is different from duty and share".

The Iraqi side expected that the Iraqi right to the royalties would be self-evident, and that the difficulty was to be found in the more basic issues connected with the structure of relations, i.e. the questions of participation and making basic decisions. But, after speaking about seriousness in deciding the issues, Mr Stockwell declared: 'I cannot say that we shall pay royalty dues before we enter into negotiations which we hope to finish once and for all". This meant that the companies would not pay to Iraq its financial dues except in return for solving the other problems, i.e. the companies would use their payment of dues as a means of pressure and bargaining power to make the Iraqi side more "sensible" in discussing the claims of the companies".

It was natural after this first exchange that the head of the Iraqi delegation should start a counter attack: "We insist on discussing the question of expending royalties first.... "It should be obvious that we have not come to the negotiations pushed by a desire to discuss and exchange opinions. We want to get our rights and decide all the issues. Certainly our relations with the companies after the negotiation will be on a new footing".

Before we look at the other sessions, we should look into the question of the previous offer by the companies in 1964, concerning royalty expenses. The companies, in their negotiations with OPEC in that year, specified certain terms for Iraq without which royalties cannot be paid. One of these terms is compulsory arbitration, which was in fact a way to defuse Law No. 80, by classifying it as a dispute which calls for arbitration.

Despite its significance, Law No. 80 is only one example that may lead to other examples. The basic question here is: do these companies, working on Iraqi territory, come under the sovereignty of the host state or not? If the answer is yes, then the state can issue any legislation it wishes, which must be binding to the companies, irrespective of the nature of these regulations, and no independent state allows her legislations to be subject to approval by others. What is the sense, then, in the companies' demand to have international arbitration as binding in settling conflicts? From the legal point of view the question is decided. The International Court of Justice had previously refused to look into a claim by the Iranian Oil Company against Iran when the Nationalization Law was issued at the time of Musaddaq. The International Court of Justice replied that it was not within the powers of the Court to look into complaints between a state and a company, and that local courts alone had jurisdiction over such matters.

Why should the companies in 1972 insist on compulsory arbitration, which they provided for in the agreements of the first concession, and retained in the February 1952 agreement? It only means they were insisting on retaining the original structure of relations, and retaining the status of an independent state dealing with the Iraqi state, instead of coming down to size, that is, as a company operating within a state.

Another condition which the companies wanted to enforce in return for approval of the payment of royalty expenses was that the operating companies should be given favourable treatment. This means that Iraq has to undertake to prefer the operating companies with their traditional concessions over any other oil company in all matters connected with oil, thus depriving Iraq of new possibilities in the world of oil and keeping the country under the economic control of the cartel companies.

These terms were refused by Iraq when they were first submitted in 1964; now the companies in 1972 said they were presenting the same terms once more in return for their condescension of the approval of the payment of royalty expenses. Out of magnanimity, Mr Stockwell said it was possible to rephrase the previous offer in order to make it more acceptable to the government.

In the second session, the companies insisted on compulsory arbitration and the favourable treatment. But the Iraqi reply was: "When the companies undertake to settle the question of royalties and drop their conditions, then the question of arbitration can be discussed among the other points. Apart from that, there is no point in continuing the negotiations. We have not yet seen any progress in these negotiations". And the session ended.

The negotiations could have ended with that session. But it seems that discussions behind the scenes made a breakthrough where the way had seemed completely blocked. Undoubtedly each side wanted to know what the other had up their sleeves, so it was necessary to continue.

The companies sent a letter dropping the condition of favourable treatment but insisting on arbitration, though limiting it to the question of royalties. But the Iraqi delegation stated that the question of arbitration was a question of principle, and whether arbitration was limited to royalties or any other question, it impinged upon the sovereignty of the state. Therefore, after studying this letter, it was decided not to accept the amendment mentioned. It was not reasonable for Iraq at that time to concede to conditions concerning arbitration. "We believe that change must be in the right direction."

In point of fact, the concession which the companies made about the favourable treatment was nothing but a formality, since the negotiations were going on at the National Assembly, and since INOC was actually at the time exercising direct development and dealing internationally with all the independence which would secure Iraqi interests. The term had been practically invalidated already, and open confrontation over this point would have been useless.

At this point there was nothing against continuation – though intentions were becoming clear and beating about the bush was apparent. The Iraqi delegation presented the government claims, and Mr Stockwell asked to be given some time to go to London to study those claims with the companies and present a composite offer from those companies.

Nothing against that either... but "the companies have to realize that when we agreed to enter into these negotiations, the political leadership had carefully studied this matter, and we are not ready for another procrastination. We want to decide upon these questions without delay – either to have new relations with the companies or there will be no relations and no companies either; each to decide on his own". Thus ended the fifth session on 19 January 1972.

By 1 February, the delegation from the companies had returned from London. Stockwell presented his composite offer, which may be summarized as follows:

1. The companies claim compensation for economic damages done to them by the legislation of Law No. 80. The compensation is to equal 12.5% of total oil produced by INOC.

2. Of the oil produced by INOC, the companies want Iraq to approve sales of 1.1 billion tons in a period of 20 years at a reduced price of 1.62 dollars per barrel.

3. The companies are willing to form a special company with headquarters in Iraq, with the government represented in this company's Board of Directors, which is to manage the operations of the three companies working in Iraq. Those companies are to retain control over three areas: operations programmes, basic budgets, and other questions connected with the Government.

4. The companies agree to the payment of royalty expenses and respond to the Iraqi claim concerning amendment of marketing expenses in accordance with agreement concluded with OPEC countries.

5. The companies agree to raise production in the south to about 45 million tons.

6. The companies agree to pay a total sum of ten million dollars for the rest of the Iraqi claims.

The companies also stated that these proposals must be taken *in toto*, and no point may be taken separately from any other point.

The proposed compensation stems only from the imaginary claim that the companies own the territory and wealth of Iraq. The claim of 12.5% of total oil produced by INOC is equal to the production due to the Iraqi Government from total oil produced by concessionary companies – as royalties in the capacity that Iraq owns the land and the wealth. That is the first inconsistency.

The second inconsistency concerns the price proposed by the companies to buy INOC oil at 1.62 dollars. For the sake of comparison, these were the agreed prices of Gulf Oil (before adjustment caused by the decrease in the value of the dollar which increased prices by 7%): \$2.155 (1971), \$2.59 (1972), \$2.3065 (1973), \$2.474 (1974) and \$2.586 (1975).

The attitude of the Iraqi government was to lead on and corner the companies. Their arguments may be summarized as follows:

1. Iraq completely rejects the principle of compensating the companies for the legislation of Law No. 80.

2. Iraq must be compensated for the decrease in production and damages caused by the companies to the Iraqi Treasury during the previous period, and interests on sums pending with the companies must be paid.

3. The Iraqi side agrees to sell limited quantities of oil for ten years, provided such sales be at prevalent prices and on commercial terms, without reductions.

4. There is no use in forming a special company, where the Board of Directors of such a company would have no powers to look into financial matters, planning and relations with the state. We demand that the company's administration move to Baghdad, and that the state participate in the Board of Directors and supervise the accounts.

5. The Iraqi side insists that royalties be paid according to state prices plus amounts of interest due.

6. The Iraqi side insists on an increase in production in the south and north fields, in accordance with a programme taking the increase to a total of about 160 million tons in 1975, in both areas.

7. The Iraqi side insists that all claims and problems which Iraq has with the companies be handled one at a time, and not *in toto,* as the companies want.

8. Iraq cannot be lenient about the protection of oil wealth, since the companies have employed methods of production which save the

companies some expense, but cause at the same time economic damage to the Iraqi oil wealth.

They concluded:

"The negotiations are not a seminar where we can be educated to the ideas of others. Either we decide the problems and reach a complete solution, or we see no point in the negotiations at all. We consider the signing of the concession to have been effected between two unequal sides, between powerless rulers and powerful companies. When we now demand our rights that were stolen from us, the companies should not stick to the spirit and mentality of the past. They should take into consideration the present government, the nature of its leadership, the development of Iraqi interests and the effect of the presence of the companies in Iraq... and also, the nature of the development that has taken place in oil relations on the international level. We have a common proverb which says: 'Do not throw a stone into the well you drink from'."

It was necessary to remind the companies' delegation once more of the general political framework which surrounded the negotiations. The political reality alone could make this delegation change their proposals, and narrow the difference between the two sides.

A few years have passed since these negotiations took place and we all know the patriotic forces completely succeeded. So, there is no harm in remembering the interesting aspects, and Mr Stockwell was really interesting. He was an icy being impervious to any shock. After the previous sparks, he resumed talking normally in his previous tone as if nothing had happened.

"Concerning the payment of royalty expenses, we have previously said..."

Sarcasm could not touch him either. When the chief of the Iraqi delegation said during the debate: "According to this, we cancel Law No. 80, and levy taxes from the companies! ". Stockwell answered: "Sorry, this is not what I mean," and continued, ignoring the sarcasm.

No use! Therefore, the final word was: "We categorically refuse the proposals presented; they cannot be accepted in any way. Unless a new offer is presented by the companies there will be no point in continuing the negotiations".

In the penultimate session, the new offer from the companies was discussed. The offer was made on February 5, and the session was held on February 9. There was nothing new in it except raising the purchase price of the INOC 1.1 billion tons of oil from $1.63 to $1.65.

## NOTES

For information on negotiations, see the text published by the Iraqi Government under the title: *File on the Battle with Oil Monopolies – Negotiations and Nationalization. Complete Text of Minutes of Negotiations with Oil Companies.*

CHAPTER 9

# FROM THE NEGOTIATING TABLE TO THE STREET

On 8 December 1941, Sir Winston Churchill, then British Prime Minister, sent a letter to the Japanese Foreign Minister saying: "Instructions have been issued to the British Ambassador in Tokyo to inform the government of the Japanese Empire, in the name of the British government, that the state of war has been declared between our two states". Churchill concluded this letter by saying: "I have the honour, sir, with full appreciation, to be your humble servant." In his memoirs, Churchill mentioned that "some people did not like this courteous expression; but there is nothing to stop you from being polite to someone whom you want to kill personally".[1]

It seems that Stockwell believed in this wisdom. Before he turned his back to leave the hall, he said at the conclusion of the eighth session of the negotiations: "I would like to express my sincere thanks and gratitude for the understanding which you showed in these sessions". And definitely the man did not mean any of the sweet words he said. IPC had already decided to exercise a ferocious economic pressure against stubborn Iraq.

In March, the companies began a sharp decrease in production rates. Immediately, the Ministry of Oil and Minerals warned the Follow-up Committee. Instructions were issued to ban publication of this news and to follow up the situation closely. On April 29, the Ministry of Oil began to inform the departments concerned about the wosening situation. The Minister of Finance began to warn of the necessity of taking immediate measures.

> Confidential letter No. S-526 dated 9 May 1972: To: The Chairman, Follow-up Committe on Oil Affairs and Implementation of Agreements:
>
> The Ministry of Oil and Minerals notes serious developments in rates of oil exports from northern fields, during the last two months [March and April]. This fall is having a serious effect on assessment of government revenue from oil. If this fall in production is to continue, it will cause a decrease in returns ranging from 50 to 86 million pounds for the present year.
>
> The Ministry of Finance finds it imperative to emphasize the seriousness of this sizable decrease in óil returns, and its effect on the financial condition of the general budget, and the monetary

condition of the Treasury, since the budget was approved at a deficit. If the oil companies do not go back to the former rates of export, this Ministry finds it imperative to revise all state-expenditure, and to curtail the same to face this situation immediately.

If our proposal to revise state-expenditure is to be approved in principle, we shall present detailed proposals to achieve the same. Kindly advise us of your opinion on this subject.

Minister of Finance

The Command received the reports of the Ministry of Oil and the suggestions of the Ministry of Finance with due attention, but without panic. It was the war for which the command was prepared for some time. It was decided to postpone the publication of this news until preparations for general attack were ready. At the same time, there was a need to retain the initiative and not to lose sight of the goal.

In reality, each of the two delegations round the negotiating table had been but the visible tip of the iceberg. The ship would founder and sink if she were not careful to avoid the deep and wide bulk of the iceberg under the surface. The same thing happens to anyone who follows the oil situation and tries to understand what goes on, judging merely from the picture the negotiators present and from the conversation going round the table. Every word or allusion or manoeuvre in that hall was the result of careful calculations made by each side, both of his powers and the powers of the adversary.

Before and after each session, the companies' delegation sent reports, and sometimes representatives, to the IPC headquarters in London. With the important developments, the London office held extended meetings and communicated with the governments concerned in Washington, London or Paris.

The same thing happened in Iraq: The Follow-up Committee closely watched what was going on in the negotiation hall, and, after discussion, instructions were issued.

If we have not yet become familiar with what went on in that period behind the closed doors of the IPC London office, we have tried, as much as possible, to learn the inside story on the Iraqi side.

We start here with some statements by one of the Party leaders in Iraq.

"There are systems, whose constitutions provide that a 'certain party' 'a certain organization' is the leader, and the source of authority. But everyone knows that 'party' or that 'organization' have no authority outside the name and appearance, and that the department of 'Public Security' for instance may possess much more authority than the 'party' or the 'organization'. This is a common feature, emphasized by some situations existing in the Arab world. For instance: the military establishment, which led or particiapted in some changes, may in some systems play a part much greater than the part announced. And this part is certainly at the expense of the role of the party or the organization in whose name the rule is exercised. In addition to the military establishment, there are tribal, sectarian, family or other types of relations, attacked in the press and official communication media, but which play in fact a vital part in authority and which may reach a serious level in some cases.... It is common among systems which claim allegiance to a party or an organization that the headquarters of that party or organization, on official occasions, becomes a hive of activity for responsible officials, people and communication media. But when there is real work to be done and decisions to be made by the authority, attention should not be directed to that headquarters, or to those who sit behind its luxurious desks, but to those who have the real authority, in making decisions and managing affairs, probably without cameras or loud-speakers.... During four full years of scrutiny both on normal and on decisive days of thousands of practical examples of small and large decisions, from the uprising of July 30 until the nationalization of oil on June 1, 1972, I believe that the majority of people in Iraq, and the majority of Arabs and foreigners who are familiar with facts, have become reasonably convinced that the Party establishment organization and personnel exercise the jobs they claim in actual fact, and that they are definitely not a façade for other powers."[2]

These words were published under the title "Revolution of the New Road" on 17 July 1972. I have quoted them in order to move to the inside story of the Iraqi team conducting the oil battle.

These words touch upon something of great importance, of great significance, in the setting up of a revolutionary state – or in fact in the setting up of any state – and that is the question of the validity of the political institution and the extent of its control over the other institutions of the state and society, and ultimately the method of making basic decisions. Efforts have often faltered in our Arab world in the attempt to reach a politically sound version, in this respect, after the fall of the traditional institutions and systems.

This question has to be kept in mind while talking about one of the most serious decisions, and how it is implemented. So, how far has the Iraqi Revolution succeeded in exercising the sound version, which was previously missing?

In the summer of 1970, lengthy meetings were held by the Revolutionary Command Council and the Regional Command to evaluate the results of the unsuccessful secret negotiations with the concessionary companies. In those meetings the oil policy for the next stage was drawn up, and the Follow-up Committee was entrusted with its implementation. The course of the policy advanced towards a general aim of nationalization. That action was the submerged four-fifths of the iceberg supporting the Iraqi negotiating team, and that was the basis of the Nationalization Law of 1 June 1972.

After the preliminary reports to the Follow-up Committee about the development of the 1972 negotiations, the political Command discussed the attitude of the companies. Although the discussions in those meetings are not published, yet it may be said that they were about whether preparations for nationalization were enough advanced, about various alternatives for the method of confrontation and the version of nationalization in case it was found possible at that stage. The conclusion was that nationalization was possible, and the Follow-up Committee was charged with concentrating efforts in order to finish all preparations at the earliest stage possible, so that no one outside the political staff concerned should know of the nature and direction of those preparations.

The companies, as usual, tried to prolong the negotiations. This by itself was a kind of test of power. If Iraq agreed to procrastination and lengthy negotiations this would show that Iraq was not ready for another method of confrontation. The instructions of the Chairman of the Follow-up Committee were to stop any attempt at procrastination and scheming.

The companies also tried during the negotiations to be strict on minor questions, to appear as if they were making great concessions if they had to change that attitude, and to ask in return for a concession from the Iraqi side on more serious questions, like the Law No. 80. The instructions to the Iraqi delegation were very clear: the categorical refusal of all of these methods to show that the negotiations were not traditional this time, and that they must be carried out on the basis of the realization that the structure of relations must be radically changed in accordance with the change in the balance of powers, and that negotiation and bargaining was only possible within the new structure.

The companies thought that the Iraqi national authority was over-estimating its power, and that the balance of power had not yet change in the manner imagined by the Iraqi side. Therefore they refused to change the structure of relations, and took decisions aiming to restore "realism" to the Iraqi mind, and thus agreed to end the negotiations. But the companies realized that if the balance of power had not changed yet, then it was well on its way to that change. Therefore, it was not enough to end negotiations, but punitive measures had to be taken to jeopardize any measures that may be already underway in Iraq. The rate of oil production in the northern fields began to go down sharply in March 1972.

The Command was prepared for this possibility since the decision about direct preparation for nationalization was issued in January 1972.

> "Preparations were started to direct the national economic and financial resouces according to new plans, taking into account the possible results of the development of relations with the companies, and in accordance with the calculations of the Command and their insistence on not making concessions about our legitimate claims. These preparations set us on the path to unilateral legislation if the companies insisted on their attitudes."[3]
>
> "Since the early months of 1972, the political Command had asked the planning offices to draw up two development programmes: one to suit the minimum level of expected resources, in case these resources fell short of levels assigned by national development plans, and the other to use the existing natural resources, which could ultimately fulfil the entire aspirations of the political Command."[4]

An Iraqi official says,

> "All calculations were made on the basis of the 'worst possibilities'. This expression was recurrent in all our publicity systems in the period following nationalization. The worst possibilities did not mean simply a decrease in government oil-returns, but it meant the possibility that the Government could fail to export one drop of oil for two years, whether from Kirkuk or from Basra (and Rumaila oil was limited at that time)."[5]

The Government at that time relied on the readiness of the people to sacrifice, that is, to be able to live on half salary for two years, as

Saddam Hussein said seventeen months before nationalization. Besides this readiness to sacrifice, the state directed financial resources, as previously explained, so that a portion of foreign currency was sent apart to provide for minimum necessary imports to hold out for two or three years – as was stated by a responsible person at the time – and during the two years, the revolution could, in one way or another, break the siege which the 'worst possibilities' envisaged in the first stage.[6]

In fact the Revolution was hoping to overstep those worst possibilities, which they they also studied and prepared for, so that there would be no surprise of any sort. On the practical side, the political Command and the Follow-up Committee had secured semi-certain possibilities for marketing nationalized oil for some time.

In any case, IPC started reducing pumping rates from Kirkuk fields, giving, as usual on such occasions, innocent economic reasons. The claim was that the decrease was due to a similar decrease in shipping fares which meant that oil going short distances became unable to compete with oil shipped for long distances, such as that going to Europe from the Arab Gulf around the Cape of Good Hope. Thus the Gulf oil reaching Rotterdam, or even the south Mediterranean, would cost less than oil exported from Banias to these places. But the Iraqi officials noticed a discrepancy in the argument. Shipping fares were continuously going down throughout the entire year, and were not intermittent or changeable. If the reason given were true, then it would entail a fall in export from Kirkuk throughout the year at the same steady rate. But it was noticed that production went up at certain times and down at others. Between January and June 1971 production went up, and 'it so happened' that this period witnessed some signs of ease in relations between the companies and the Government, as a result of negotiation about prices and the signing of the February 1971 agreement, in accordance with the Tehran Agreement, and then the signing of two agreements, the East Mediterranean, and the Real Cost Agreements in June 1971. Between July and September of the same year, the company decreased production by two-thirds. Again, 'it so happened' that this was the period before negotiations. The Iraqis saw this as a form of pressure and warning, and negotiations were postponed. Then 'it so happened' that production went up to maximum rates which export could cope with between December and February, the months of negotiations. Then suddenly after the failure of negotiations[7] the "merely commercial" circumstances called for a change in production levels and the decrease was sudden.

Fabricating excuses about the unfavourable economics of Kirkuk oil could not convince anyone who was familiar with these vacillations and their implications, even without a technical discussion of the nature of the oil market, and whether it is competitive as the companies claim, or a monopoly. Yet, it is useful here to indicate that the companies' insistence on claiming economic reasons for decreasing Kirkuk production was not simply for the sake of excuses. The companies also monopolize the "mysteries" of marketing. And when data and information leaked out or were announced, it was only to create perplexity among the Iraqi ranks. When suspicion was stirred up about the existence of a real problem in Kirkuk oil marketing, hesitation was expected to rise: how could Iraq take a radical step and embark upon exporting to the international world jungle of which she knew nothing, when the companies with all their experience and potentialities were faltering? The companies were waging a psychological war when they decided to decrease exports. And the Iraqi reply? The economic and financial measures including preparation for nationalization which started a long time before decreasing production, as was previously explained.

The other measure was taken by the Follow-up Committee charging the Ministry of Oil and Minerals to send for the representatives of the companies to study the situation. The Ministry showed that the sharp fall in production rates and the serious financial results thereof on government returns can not pass unheeded. The Ministry added that it the companies consider the decrease a *tour de force,* then a serious crisis would follow.

The companies answered that they appreciated the Government's anxiety. After much give and take, the companies came up with a proposal which was the ultimate in provocation. They declared that the companies' production of the northern fields would continue within 30 million tons if the reasons justifying decreasing production should continue. But they were ready to increase production to 50 million tons if the Government agreed to decrease the price per barrel by 35 cents. This proposal simply meant that Iraq should agree to play a dirty part in destroying the average price of oil, and destroying the price structure which the OPEC countries had endeavoured, after great insistence, to reach and preserve. The companies asked Iraq in particular to play this role, knowing very well the distinguished part played by Iraq in developing OPEC and supporting its unity. It meant that the companies wanted in one blow to destroy the esteem of the Government in Iraq, and with it the unity of the OPEC countries.

The Follow-up Committee naturally rejected the proposal. The Ministry of Oil informed the companies of this rejection. But since the affair had become a *tour de force* rejection alone was not sufficient. Acting upon instructions from the Command, the Government presented three alternatives to solve the problem.

The first was that if the companies were not willing to produce and export more than 30 million tons, it would be possible to solve this problem by letting the companies produce 57 million tons (which was almost the top capacity of the northern fields), of which the companies could take the 30 million tons which they claim they could not exceed in marketing, and the Government would take the other 27 million tons to undertake marketing for the Government's own account. The Government would naturally pay production and shipment for that amount.

The second alternative was that the companies would not produce and export more than 30 million tons, this meant that a large part of the northern installation capacities – especially pipeline capacity – would not be in use, since the total capacity of these installations was about 59 million tons. The Government proposed to personally use this surplus capacity, since installations and establishments set up by the companies in Iraq – whether in the North or in the South – were financed by both sides, i.e. Iraq paid 55% of the cost of these installations. It would be possible to conclude a long-term agreement between the Government and companies, whereby the latter limit the size of pipeline capacity they want to use, and forgo the capacity not needed, which the Government would use in return for payment of the share of the cost for using such installations.

The third and most radical suggestion was that if oil production from the North was not profitable to the companies, while southern oil exported from the Arab Gulf was more profitable, then the companies could hand over the northern fields to the Government, and replace their loss by increasing production in the South. The Government, on its part, agreed to the development of southern production from fields at the disposal of the companies in such manner as to compensate for oil lost or relinquished by companies in the North. These southern fields were sufficiently large that they could increase production to the level that could compensate the companies.[8]

The Ministry of Oil and Minerals explained to the companies that they were "free" to choose any of the three alternatives by a date not later than 23 May. The companies agreed to study the proposals.

But to study what? The Revolutioary Command had recovered the initiative by these proposals, which were not considered tactical counter-attacks, since they did not deal with decreases of production or otherwise. The proposals were a move towards a general strategic attack. Decreasing production was no longer a problem. On the contrary, the state accepted and used this issue in order to go back to the orgins of the problem – changing the structure of relations between the state and the companies, and moving the state into the position of control in the oil industry.

The news of the production decrease during the three months from March to May was still a secret, and all communications took place behind closed doors. But, with the obstinacy of the companies, and the refusal of the Command to be on the defensive (which was obvious in those three alternatives) the situation could stand no waiting. If the Command were determined not to lose the initiative, then to continue with the plan was a necessity.

A joint meeting was held by the Revolutionary Command and the Regional Command at that time, and the meeting issued a number of conclusive decisions:

1. Zero hour: June 1. That is, the nationalization decision was to be on the first of June.

2. Declaring comprehensive popular recruitment and emphasizing unity among patriotic forces.

3. Enlarging the powers of the chairman of the Follow-ip Committee. To run a war demands a central administration of operations, and the means to make immediate decisions to face successive and emergency changes.

To carry out the third decision, the Follow-up Committee had to submit data and studies, and Saddam Hussein was making decisions in the light of constant consultations with President Al-Bakr. Within the framework of these decisions, the battle with the companies was carried to the street. Instructions were issued to the Ministry of Oil to make a statement about the pressure exercised against the Iraqi economy. On 14 May a statement was made to the public, that the fact the companies had decreased oil production rates in February and March was expressed for the first time, and that in the first week of May the rates were really low. The statement said that, "The objective of the companies to exercise pressures on the Revolutionary Government was obvious to whoever is familiar with production affairs, and their connection with the negotiations that are going on between the Government and the companies".

On 17 May a statement by the RCC announced a greater acceleration in confrontation, by addressing an ultimatum to the companies:

> "Early in 1972, the Iraqi Government entered into negotiations with the group of foreign oil companies operating in Iraq, in order to decide all pending questions which form the legitimate rights of our people, and in order to define future relations, based on respect of national sovereignty and on dealing from new positions. The policy of causing harm to the Iraqi people will only lead the companies into serious consequences, and this policy will not affect the Iraqi fortitude as much as it will lead to serious effects on the interests and future of those companies, not only in Iraq but also in the entire Arab world....
>
> "We address an ultimatum to these companies and demand:
>
> "1. A quick response to the Iraqi demand to raise production rates to the utmost capacity of pipelines, and immediate measures to reflect this tendency.
>
> "2. The companies should undertake to agree with the Ministry of Oil and Minerals to lay a production programme, permanent, long-range and with a scientific basis, and to take positive initiatives to achieve this aim.
>
> "3. The companies should present a positive offer about claims presented by the Government negotiation team during the last negotiations.
>
> "To look into this ultimatum and claims, we give the companies a period not exceeding two weeks with effect from the announcement of this statement... Otherwise, the Iraqi Government will see no option but to take all legal and legislative measures deemed necessary to protect the national interests and legitimate rights of our people".[9]

What is noted in the statement is the following:

1. The Command did not wait for the companies to answer, but it brought the masses into the battle.

2. A commitment before the masses that the issue is not limited to production decrease, but to the entire question.

3. The short period of the ultimatum.

4. A declared committment that the State will resort to a unilateral decision, i.e. legislation, in a period not exceeding two weeks.

After issuing the ultimatum, an RCC decision announced austerity measures "taking circumstances into account and safeguarding against every emergency to save the course of the Revolution". The decision provided for complete austerity measures in all aspects of government expenditure including the cessation of the development programme of 1972/3 and austerity measures concerning expenditure of foreign currency assets belonging to Iraq, in accordance with instructions issued by the Central Bank.[10]

It may be asked here: why declare the intention and even the date of a legislation to be issued by the state? Does not this declaration sound like wasting a chance to surprise the adversary, knowing that surprise and trickery are needed in every war?

The answer is that the idea of resorting to legislation was not a secret, since it was referred to in the warning of the Iraqi delegation more than once during the negotiations. And, talking about legislation in that statement was not upsetting any surprise plans, since there was no room for surprise or manoeuvre. On the other hand, acceleration towards comprehensive confrontation necessitated the recruitment of the masses by including them in the situation and its ramifications.

But the Command, despite this declaration, was hiding the real element of surprise and manoeuvre. The principle of resorting to legislation was known, but the content of the legislation was a hidden secret, and all communications and publicity – throughout the period following the ultimatum – endeavoured to leave this an open question. In fact the leadership of the Follow-up Committee resorted to camouflage to take attention away from the possibility that expected legislation may reach the level of comprehensive nationalization. Instructions were issued to the Ministry of Oil and Minerals to prepare three sets of legislation:

1. Draft Law for exploiting surplus energy.
2. Draft Law for limitation of minimum production.
3. Draft Law for curtailing concession to 25 years, instead of 50.

The Ministry of Oil prepared the three drafts quite seriously and submitted them to the Follow-up Committee. The Ministry technicians did not suspect that those drafts were in fact on their way to be field in the Committee archives. But, before this was done, the Command saw to the leaking out of those drafts to the companies, as part of the camouflage of the real plan: nationalization.

## NOTES

1. *The Memoirs of W. Churchill* vol. 2, Arabic translation by M. Shalbi (Cairo, 1970) p. 209.
2. T. Aziz, *op. cit.*, (chapter on role of Party organization).
3. Saddam Hussein, *op. cit.*
4. S. Al-Shaikhly, *Oil Nationalization and National Development* (Baghdad, May, 1975).
5. T. Aziz, conversation with the author.
6. Letter from Baghdad by E. Rouleau to *Le Monde*, June 3, 1972.
7. F. Al-Chalabi, *Developments and Oil Issues in Iraq*, T.V. interview on May 17, 1972.
8. S. Hammadi, in T.V. interview above.
9. For text of statement, see *Negotiations and Nationalization*.
10. *ibid.*

## CHAPTER 10

# COUNTDOWN

After the ultimatum, the countdown began. Every day drew zero hour closer. The pages and leaders of *Al-Thawra* daily counted the tempo.

On 18 May, the headlines read:

A Final Warning to the Oil Companies. We shall undertake what secures our interests if the companies do not respond to our demands within two weeks.

On the same day, the paper started to publish the first series of the complete text of the negotiations minutes of February 1972, an unprecedented procedure. This took part in accelerating the popular recruitment against the companies.

On 19 May (13 days from the deadline:

The leader read: On the Economic Decisions Taken by the RCC – economic, political and psychological strength of the home front in facing monopolies.

On 22 May (10 days from the deadline):

The leader read:

On the initiatives of Labour Unions and State Departments – The Masses' sense of responsibility is our way to victory in the patriotic battle.

On 23 May (9 days from the deadline):

The leader read:

Our battle with the monopolies is a patriotic battle which needs participation from every citizen.

On 24 May (8 days from the deadline):

The leader read:

On the honourable patriotic attitude of the Iraqis working with the oil companies: dedication of the people and calculations of the monopolists and imperialists.

On 25 May (7 days from the deadline):

The leader read:

The battle with monopolies is one aspect of the battle with the imperialistic-Zionist alliance.

On 26 May (6 days from the deadline):

The leader read:

When the decisive moment arrives, the Party and Revolutionary Command will fight the battle to the end.

On 27 May (5 days from the deadline):
The leader read:
The first of June shall witness the victory of the people over the monopolies.
On 28 May (4 days from the deadline):
The leader read:
The open manoeuvres of the companies will not deceive us: our demands are an indivisible whole.

The last leader deserves a closer look. It is concerned with some news which leaked out through company channels about the return of Kirkuk production to previous rates. *Al-Thawra* leader answered decisively to this move.

"The latest company manoeuvre is to try to present the battle as if it were concerned with the increase or decrease of production and not as a battle defined by the RCC's historic and decisive statement..... Whoever understands our language knows that our demands are not limited to increasing production to the top capacity, since this is only one of the issues included in the historic ultimatum. The result of this move on the company's part is that it accelerated the measures that the Government was intending to take. For their response to be real, it has to be complete, without side-stepping the three points mentioned by the RCC statement. Otherwise, any partial response to our demands is a manoeuvre, but this time it is an open manoeuvre and does not display any degree of intelligence."

Then the article wondered why the companies should manoeuvre, and said that they were trying to divide up Iraq's demands. Through fear of popular opinion, and after the ultimatum, the companies try to create perplexity and influence international solidarity, especially within the framework of the OPEC countries and the Arab oil-exporting countries. "These manoeuvres of the companies are self-exposing... and the psycho-political aspect is very obvious in the behaviour of the companies". When the Government insists on the demands, the companies decrease production; when faced with insistence and challenge, the companies pretend to retreat and follow outmoded side-stepping methods. "Is this commercial behaviour, or is it a psycho-political attitude? ".

"From now on London shall not decide the size of production from the fields which we possess, and London shall not decide on the issues which concern our national wealth. Baghdad shall decide, first and last. The question is more than production going up or down. It is a question of patriotic and national dignity. We are resolved to fight the battle to the end, and we shall win."

On 29 May (3 days from the deadline) *Al-Thawra* had these headlines:

Saddam Hussein: The strength of the home front is more valuable than any hard currency in the world, stronger than any weapon in fighting colonialism and international imperialism. The first of June is Victory Day... a day when oil monopolies shall kneel before the rights of our people.

The leader said:

On the call of Iraq for an emergency conference of the Arab oil-exporting countries. The Arab countries should prove their strength as a group able to face up to the monopolies.

On 30 May (2 days from the deadline):

The leader read:

Our battle with the oil monopolies prepares the way for a comprehensive uprising in the Arab world.

On May 31 (1 day from the deadline):

The leader read:

Tomorrow the monopolies shall find out for sure that we mean every word we say.

June, the ultimatum is over

The leader read:

The first of June... the day of the people's victory over the oil monopolies.

*Al-Thawra* daily was not concerned only with exciting enthusiasm during the two weeks of the ultimatum. When the RCC issued the ultimatum to the companies and decided the austerity measures, the enthusiasm was already widespread. But it is also fair to say that some measure of hesitation was felt in some quarters. After long years of erring leadership, and moods of defeatism at times, wondering was legitimate: was there enough preparation? Could the leadership sustain this attitude to the end? .

There is no doubt that the successive articles in *Al-Thawra* helped to disperse these hesitations. Political awareness was at a peak in the party organizations of the ABSP.

But the extensive popularity, this superb publicity effort, was backed up and headed by the Follow-up Committee. Saddam Hussein and his companions hardly left the National Assembly headquarters. From the Command headquarters decisions were issued in every direction for general preparation and co-ordination of movement, in addition to action in various directions from the ministries and departments concerned: the Ministries of Finance and Planning and the Central Bank – in addition to Arab and international communications.

On the day following the announcement of austerity measures, for instance, and amidst wide popular response, the assistant to Saddam Hussein in the Follow-up Committee was submitting to the Command definite suggestions and recommendations. This is a sample of the reports of that period and the system of inside action at the headquarters of the Follow-up Committee.

18 May 1972
To: The Vice-Chairman, RCC
Dear Sir,

1. To safeguard against potential anxiety in commercial quarters, instigated by companies' agents to disturb the internal situation by encouraging citizens, through pressure and rumour, to draw their financial assets and deposits from the banks, I suggest drawing the attention of the Controller of the Central Bank to this question. I think that he should prepare confidential instructions to meet the situation, by personally watching the daily withdrawals from the banks.

2. Rumours are expected to emerge and interpret the paragraph in the announcement of the austerity measures about stopping circulation of foreign currency as if it is meant to stop the import of consumable commodities. This may encourage some citizens to rush into purchasing such commodities. This possibility calls for the Ministry of Economics and Commercial Establishments to watch the situation closely and study the measures that must be taken to stop such phenomena when they emerge.

3. Cessation of work on projects financed by the Plan or the budget may freeze some of the contractors and financiers. Since you have previously suggested studying the possibility of directing this capital to other fields of production, I suggest that the Minister of Planning or another person concerned issue instructions or decisions and offer suitable facilities to contractors to employ their

capital in constructive works and public projects in the Arab Gulf areas, and with definite assurances, for fear this capital may escape from Iraq definitely.

4. Since I am in charge of the professional bureau at present, I have asked the central bureaux officials to stimulate the enthusiasm of the professionals and technicians, and make their participation in this battle a real one, by spreading the idea of the necessity of supporting the state in the battle against the companies, through announcing their readiness to forgo their allowances or part of their concessions, to put that in the service of the state when it is found necessary. The success of such a campaign needs to come from the independent patriotic forces. The first positive indication in this respect came in a unanimous telegram from the engineers to the RCC. The psychological state of the citizens is very high. The RCC decision has been received with great and active consent by various groups of people, which makes it imperative for the publicity bodies to sustain this enthusiasm and instruct the citizens in the causes of this battle every day, until the end of the ultimatum, in order to prepare the people for the post ultimatum period, so they will be able to accept the decisive revolutionary steps and be ready for sacrifice with a high morale.

5. And here also comes the role of the organizations, and especially the Economists' Association. Therefore I suggest to prepare for a high level seminar, to explain the latest developments with the companies, and to explain the decision to take austerity measures, and to call upon the citizens at the same time to do their duty as a people, fighting for their sovereignty and independence.

The Lawyers' Association can play a similar role by conducting a seminar to take up some of these questions.

To add to popular support, I suggest inviting some of the citizens concerned with oil affairs to a special seminar, conducted by the Minister of Oil and Minerals, to air their opinion on the situation, and to make them appreciate the seriousness of the battle, and the necessity of their participation. An outline of such opinions may be published in the newspapers if the situation is favourable.

Awaiting your advice.

Such suggestions and recommendations, as a result of communications inside and outside the system, were inspected by the chairman of the Follow-up Committee. He would refuse some, amend other suggestions or approve them in the submitted version. In all cases his comments become decisions for various quarters to carry out immediately.

In the previous report, for instance, the first suggestion became a real decision directed to the controller of the Central Bank. The second suggestion became a decision directed to the Minister of Economics, as well as to the publicity and Party systems to use in their political awareness campaigns. Concerning the third suggestion, Saddam Hussein wanted to make sure that the projects stopped were actually the projects that were not yet started. "But perhaps in the future we may have to take some measures; the Minister of Planning is to be asked to study these possibilities and the necessary treatment". Concerning the fourth suggestion, the chairman of the Follow-up Committee approved the continuation of work in that direction. Of the fifth recommendation, the secretariat was asked to make the necessary contacts "to provide speed and overlook formalities".

So the ultimatum period was spent in preparation and action on all levels. In accordance with calculations based on the worst possibilities "the State set apart some of the foreign currency assets" as we have mentioned. But, expecting the worst of the worst possibilites, it was necessary to secure those assets against dangers of freezing. So, "in the days preceding the nationalization, we took a decision to move our assets of hard currency in the foreign banks to other banks not under control of monopolies and countries to be affected by our nationalization, to safeguard against those companies and countries laying their hands on it, as was done by Britain and France after the nationalization of the Suez Canal".[1] Nothing was left to chance.

With action on the internal front taken care of, it was necessary to start preparations on the Arab level and on the level of oil-producing countries, at a certain stage of the countdown.

On 23 May, Iraq asked the OPEC Secretariat to hold an emergency meeting, and preparations for that meeting were immediately started.

On 24 May, a number of ministers carried verbal messages from President Al-Bakr to Arab kings and presidents. The ministers were chosen to represent various parties and patriotic factions participating in the Government. It is useful to know the text of instructions given to the representatives.

1. Explanation of the development of relations with companies since the beginning of the last negotiations this year. Underlining the Iraqi point of view on the basic issues in the negotiations.
2. Emphasizing the fact that the companies, in their stubborn

attitude whereby they ignored our legitimate rights and decreased production, which action caused financial damages to the state amounting to an annual average of £150 million, were in fact intent upon causing us damage because of our insistence on a direct oil-production policy, and implementation of Law No. 80 (It is possible to add other reasons according to the nature of each Arab country concerned and according to whatever the representative finds useful.)

3. The Arab countries concerned are requested not to compensate the oil monopolies for what they may lose of Iraqi oil, whether because of the latest policy of production decrease or in the case of relations with Iraq developing to a critical nature. The request may be worded as follows: Iraq hopes that the company may not find encouragement to increase production in friendly Arab states to compensate shortages affecting, or which may affect, that company as a result of its policy in Iraq. Estimation of other assistance is left to the Arab countries each according to circumstances, especially where production decrease does not affect budget or development programmes in countries concerned, or the offers of immediate financial assistance.

4. Emphasizing that Iraq is serious about extracting all rights mentioned in negotiations, including limiting the companies to a permanent and agreed-upon time-table of production, securing development needs, and stopping companies from interference. Otherwise, Iraq will be obliged to choose unilateral legislation to protect its interests, which may lead to a complete breakdown of operations, which would put the future of the companies in a new context.

5. To try to obtain officially-declared statements from the states concerned in supporting the Iraqi attitude. Gratitude is extended in the name of the President and the Government of Iraq to the governments and heads of states that declare a supportive sttitude.

6. To try, whenever possible, to know the nature and limits of support undertaken by each of the states concerned.

7. The above mentioned points form the main indicators for discussion. Freedom to move within these limits is left to your discretion".

After this direct communication at high levels, Iraq asked, on May 28, for an emergency meeting of OAPEC (Organization of Arab Petroleum-Exporting Countries) in Baghdad on June 7. But Kuwait wanted the meeting to be held in Kuwait City on the same date, with two points on the agenda:

1. Setting up a financial establishment to support the countries which become liable to any financial problems with the companies, especially the Iraqi Government.

2. Observing production in the member countries to see that the companies do not make use of the production decrease effected in Iraq.

The Iraqi Ministry of Oil and Minerals declared approval of the Kuwaiti initiative.

On 30 May, a statement was issued by the Eighth Arab Oil Congress which was held in Algiers. "The chiefs of Arab delegations who are following the development of the struggle, express their support of all measures taken by the Iraqi Government to protect the rights of Iraq".

On 31 May, the last day of the ultimatum, a statement issued by the OPEC headquarters in Vienna warned of the consequences of confrontation with Iraq. The statement said that the production decrease of Iraqi oil was undoubtedly a revenge measure aiming to exert pressure on Iraq to relinquish its patriotic policy and development programmes. The statement also expressed support of the Iraqi attitude in rejecting the companies' offer to revise prices of oil exported from Iraq, explaining that any attempt to cause the oil-producing countries to revise the prices agreed upon in Tehran is considered a breach of agreement articles.

What was the attitude of the companies during that period? The publicity campaign had disclosed the companies' looting, exploitation and conspiracies. It was impossible for any leadership to make light of all this. On the other hand, President Al-Bakr's message to the Arab capitals re-emphasized the state's intention to resort to legislation (news of this message leaked out to the companies). Also, it was no secret for any observer that all Iraqi departments and institutions were working incessantly. Ultimately, drawing the foreign currency assets from certain banks was an open message to the companies.

As we have said, it was no secret that the state was preparing legislation. But the secret was the nature of this legislation. In the instructions of President Al-Bakr, we find the formulation of the Iraqi attitude "emphasizing the fact that Iraq is serious in extracting the entire rights mentioned in the negotiations. Otherwise Iraq will be obliged to choose the way of legislation to protect its rights, which may lead to the final break up, and put the future of the companies in a new context. "This statement intended to leave the definite content of the legislation to be resorted to by Iraq rather ambiguous. In fact the definite content was rather ambiguous even to some of the representatives themselves.

But what was happening behind the companies' doors? What we know is that certain French quarters tried to convince the companies to make real concessions to avoid the Iraqi Government resorting to legislation. But the conversation of the Iraqi officials with those quarters shows that they had no idea that the law to be issued was a Nationalization Law.

In the discussion between the French and the other parties in IPC, the disagreement was not whether or not Iraq was to issue a legislation by the end of the ultimatum. It was about the way to face that possibility and the extent to which concessions could be made to stop it. The other parties did not agree with the French in making large concessions. If Iraq insisted on extracting these concessions by legislation, then the calculations of the London office could make assurances that this foolhardiness would destroy the economy and the Iraqi system as a whole.

Yet the companies had to make a move of some sort. To avoid stormy confrontation would be useful in any case. Even if the companies were to come out victorious after their confrontation, the consequences would be undesirable whether inside or outside Iraq. Therefore, a degree of flexibility was inevitable, together with an attempt to gain time during which the inflamatory situation would calm down. If the Iraqi side could take the bait and return to the negotiation table, it would lose face. According to this idea, meetings followed in the London office, and the technical staff were busy in drawing-up a new offer, which was to be presented on the deadline of the ultimatum. The Iraqi side at that moment, they envisaged, would be like a "drowning person" when a life-belt is thrown to him. In the last moment there would be no room for manoeuvering, and they would be obliged to accept the initiative of the companies in principle, and without much modification, as a basis for the new negotiations.

On this assumption, Mr Stockwell, the IPC Managing Director, left London for Baghdad on the morning of 30 May. On the following morning (the deadline) the man took his team to the Ministry of Oil and Minerals, and started to talk about an offer he had. But the calculations of the companies about anxious Iraq clutching at the life-belt suddenly collapsed when the talk was interrupted with a demand that the offer should be definite and in writing, to be discussed in the evening session.

The written offer which was presented in the evening session carried a real concession which would attract any drowning person. The offer provided for the companies' commitment to definite production figures which could be accepted under different circumstances:

1. Production and marketing of 1.75 million barrels/day from date of ratification to the end of 1973. The figure was to rise to 2 million barrels/day in 1974. The companies were to build up additional capacity in the installations to raise production and export capacity to 3 million barrels/day in 1977.

2. Concerning the purchase of INOC oil: the former demands of the companies were to buy 8 million barrels/day; and the new offer came down to 4 million only. The former price was $1.62 per barrel, now it became $1.65.

3. About the compensation sums which the companies claimed for land taken away by Law No. 80: the former offer said it could be paid back by obtaining 12.5% of INOC oil, but the new offer came down to 7% with a readiness to co-operate with INOC by offering facilities to increase export capacity available at Khor Al-Umayya terminal to help INOC export their oil.

4. The companies agreed about the payment of expenses and reduction of sales deductions with effect from January 1, 1964, which meant the payment of £71 million.

5. The companies had offered £10 million as a comprehensive settlement for the remaining Iraqi claims. Now the sum was raised in the new offer to £30 million.

6. Concerning participation of Iraq in 20% of capital, the companies announced their approval in principle and left details to negotiations held between the companies and Shaikh Yamani as OPEC representative.[2]

There was a real leniency in this, and the Iraqi official was right when he said[3] that it was possible for any patriotic rule to accept this offer. True, it would not be a revolutionary solution, but it would also not be doing great injustice to national interests. It was an acceptable solution if there were no insistence on basic changes in the existing structure of relations.

But the point is, that Iraq was not drowning nor in need of a life-belt. Neither was Iraq entangled in a radical trap and looking for a retreat to save face. The Iraqi leadership was planning for acceleration, and was intent upon a decisive solution. It was not acceptable according to government estimations to have less than complete control of the Iraqi state over the oil industry. When the delegation of the companies went to the National Assembly to attend the ninth and final session, the slogans and posters were all over the streets of Baghdad. In one voice they said:

"No defeatism. Forward! "

"Dignity and national sovereignty are more precious than today's loaf."

"Severest austerity at home, school, factory, office and in the government budget."

"One front, one line of struggle."

"The companies must bow before our just demands."

"Sacrifice and fortitude to liberate our oil wealth from the bonds of foreign exploiting companies."

These slogans undoubtedly represented the attitude of the Iraqi masses. But the masses in May 1972 were not directing pressure with these slogans on a hesitant authority. The authority was at the head of the masses raising slogans to face the monopolies. By the way, these are not metaphorical words. The Follow-up Committee itself worded these slogans, and others, to direct and recruit the masses.

The ninth and final session in the history of negotiations began. The chief of the Iraqi Delegation said:

> In this historic session we are limited by a definite time that cannot be changed. As a government delegation, we do not intend to prolong conversation and discussion. The Iraqi demands are clear to the companies... the Government and the people are waiting for the decisive moments that will come to an end in a few hours, and there will be no room for future sessions."

After Stockwell's explanation of the new offer, the chief Iraqi delegate said:

> "Unfortunately, we did not expect the offer to be like this. This offer does not indicate, from our point of view, a minimum degree of seriousness in reaching an agreement... We emphasize our rejection of this offer and ask the companies to present a more responsibe offer. The last date is 11 a.m. of tomorrow morning. This is flexibility on our part, and a chance we are giving to the companies.... If the delegation can not present an offer including compensation, then there is no point in meeting, and this is clear."

The situation meant that the Iraqi Command was still firm, and not hesitating to reach the brink of war and finish the negotiations. But would that Command really risk a war, or was it merely a threat to blackmail the adversary? Undoubtedly this was in Stockwell's mind when he made a final attempt to test the seriousness of this decision. After a long and thoughtful pause he asked: "Do I understand from the chief Iraqi delegate that I can ask, at 11 a.m. tomorrow morning, for an appointment for negotiations at a later date?".

The chief Iraqi delegate replied, "If the companies want to make an offer it must be presented before eleven o'clock. After that hour of tomorrow morning we shall not be ready to receive any offer, positive or negative, and there shall be no negotiations".

Why did the Iraqi side agree to extend the ultimatum to 11 a.m. to begin with? In fact, the situation with the companies was not expected to collapse suddenly. But the French ambassador asked for a last chance before making a unilateral decision. The French side showed a degree of understanding of the Iraqi situation from the start. It was politically useful, especially in dealing with the oil issue, that French interests be given a special consideration. Also, to expose the obstinacy of the companies, and prove that Iraq was ready to come to an understanding right up to the last moment was also useful to win international public opinion, and useful to strengthen agreement with Iraqi allies in the battle of nationalization.

In the light of these considerations the designer of the nationalization battle agreed to the French request to prolong the ultimatum. But the French interference did not succeed, as was expected, nor did the discussions of the companies' delegates during the night of 31 May, or in the morning; the contacts with London, all came to naught. A little before the second ultimatum, the French ambassador renewed contact with the Committee and said that his efforts failed to convince Mr Stockwell to accept the government demands.

After eleven o'clock the Council of Ministers held a meeting headed by President Al-Bakr and attended by Saddam Hussein. In that meeting the President presented to the members the nationalization decision, which was passed unanimously.

So, was the resolution, and the historic Law No. 69 of 1972, built on the most decisive and intelligent of alternatives? The legislation was a surprise in that it achieved complete nationalization which is exercising the highest level of sovereignty of the state. It was also the highest level of challenge to the companies. Leaping to this peak was making a choice of the shortest way. But the shortest way in life and in battles may not be a straight line. And the decision to attack by surprise does not mean forgoing calculations and manoeuvring.

It was expected that the decision of 1 June was to be a kind of legislation, or, say, kind of attack. But the size and direction of attack was a surprise. The companies were expecting Iraq to attack with a division, for instance, but the attack was carried by a whole brigade. Occupation was expected in some positions, but it turned to be of the whole field. But to achieve the ambitious aim with minimum casualties, it was not enough to attack with heavy troops, well trained, and the first surprise was also not enough. What was also needed was an attempt to disrupt the allied forces and to paralyse their will to take unified counter decisions.

The plan of direct attack, or the plan of nationalization, was drawn really according to these factors. This design was chosen from among a number of alternatives.

The first alternative was that it was possible to lay the plan on the basis of attack and comprehensive war in the North and the South at the same time; that is, to decide immediate nationalization of all company operations in the North and in the South.

The second alternative was to attack in the North or the South, but the attack should have partial aims in occupying some positions here and some there. This may mean to nationalize a share of the capital, and this alternative could be of various levels and forms: partial nationalization in the North (part of IPC shares), and partial nationalization in the South (part of BPC shares); or, partial nationalization in the North (in fact this was the utmost limit of some estimations); or, nationalization could be applied merely in separate areas. The proportion of such nationalization could be chosen according to political considerations. Shares of the American companies could be nationalized, for instance, as an attitude confirming the relation between oil battle and national battles against imperialism and Zionism.

The third alternative to the nationalization plan was to concentrate all attack at this stage on one of the two fortressess (in the North or in the South) with an aim of complete occupation, postponing confrontation with the other fortress to a later stage. In this case it was necessary to decide whether the North or the South was more suitable for the comprehensive attack of the first stage.

It is known now that Law No. 69 chose the last alternative, and chose the North as a field for the concentrated attack without covering BPC operations in the South. Undoubtedly this was the ideal choice, which gave the Command a chance to conduct the battle under the best circumstances. Why?

In the case of the first alternative, complete nationalization in the North and the South at one blow, the companies would be in a position of someone who has nothing more to lose or fear for. It is a position that pushes to war without limits, and to hazards usually avoided under different circumstances. The concessionary companies in Iraq were represented, as we know, by most of the international cartel companies whose influence inside the oil market could not be ignored.

A complete break up with these companies could not be achieved by a mere nationalization resolution. The Iraqi state could not merely announce a law like Law No. 69, then end the problem, and go to the world markets quite innocently, as if nothing had happened.

The problem was not whether the state had the right to issue a law, but the real problem was whether the state was able to implement that law.

Marketing the nationalized oil – in accordance with the law – in a stable manner, depended on ironing out the legal problems. The legitimacy of nationalization with the parties dealt with depended, for instance, on compensation of the nationalized capital. And since this and other problems were not settled, the dealing with the world markets remained open to pressures and conspiracies, especially when it is remembered that stable and complete marketing of Iraqi oil would not be mainly in the Socialist countries.

Naturally, if the companies chose a comprehensive war without limits, it was inevitable they would carry it through to the end. The nationalization calculations in Iraq were made, as has been said, on the assumption of the worst, that is with the supposition that Iraq would fail to make one dinar out of the oil – nationalized or otherwise – and for two years or more. This did not mean that Iraq should plan and take initiatives leading to this result. It was obvious that the Command should endeavour to minimize the complications and passive reactions. It was also obvious, after such a large and decisive blow, to leave something in the hands of the adversary that he would fear to lose. This would stir up disquiet in the enemy ranks, and would cause someone on the enemy front to advise that all available weapons should not be used, and that was what was wanted. When the companies found themseleves in a situation where they had something to lose, it would help them to swallow their arrogance, to preserve what they had, even in the hope of gaining time and regaining the initiative after a while.

The case of partial nationalization in the North and the South, the second alternative, may be said to achieve the aim. The companies would not lose everything. And at the same time, the remaining of the companies in the North and the South would insure continuation of production and marketing, and ultimately minimize the hazards of the Iraqi step. This may seem sound. But, if the objective of not resorting to complete nationalization was to disturb the adversary, and disrupt the ranks of allied forces, and paralyse their will to take unified counter-measures, then the partial nationalization in the North and the South would not achieve this aim efficiently. The companies reaction would not be different in this case from the attitude to complete nationalization, and ultimately there would be no achievement of what has been said about securing continuation of production and marketing, that is to say the step would not minimize the economic and technical hazards. Nationalization, even when partial, would make the companies in the North and the South lose their independence, and limit their power for future planning and management. In order to paralyse the will of the adversary it is necessary to leave him a real thing which deserves protection, a complete closed fortress, for a while.

To speak about the national features of the battle was not enough to prevent the backlash of the unified reaction of the international cartel companies. On the contrary. If we remember the circumstances of those days – the first half of 1972 – we realize that using such an excuse to nationalize a share from the North and the South could have had a negative reaction also on the Iraqi attitude, and an important sector of the Arab oil exporting countries was decisively against the principle of using oil politically in the national battles.

The third alternative, nationalization of a share from the North only, naturally was possible. But just as complete nationalization (in the North and the South) was skipping over some of the stages, and causing complications in the situation without need, to nationalize a share of the oil of the North was, on the other hand, less than sufficient for that stage. On the practical side, the reaction of the companies in this case would not have differed much from the case of complete nationalization of the northern oil. Why, then, should the same battle be fought, and with the same effort and sacrifice, but for a lesser aim?

The positive point about this alternative is that the blow would be in the direction of the North. In this area the conflict with the companies was acute. And even in the final offer presented on May 31, the Kirkuk production was a subject for bargaining. Commitment to high rates in production and export from these fields was a subject of controversy with the companies, because of the high cost of Kirkuk oil and the weakness of its competing position. At the same time, this logic was rejected by the Iraqi state, since production capacity of this area is 65% of available capacity of oil production in Iraq.

Therefore, if there was a chance for a blow in one of the two fields – on an economic basis acceptable to all oil countries – the target for the blow was undoubtedly the North.

The last alternative (complete nationalization of the North and complete postponement of the South) was the ideal choice. The companies – according to this choice – had an independent fortress; they had under control a real thing which they needed to see in existence. Because of this, they would hesitate before any decision taken in the battle field. And because of this care and consequent hesitation, the Iraqi side would be in a stronger position to extract the companies' recognition of the nationalization law, stronger in the discussion of settlements and compensation, and also in extracting the companies' recognition of Law No. 80. As well as this, if the companies were concerned with the continuation of their position in the South as producer and exporter, the Iraqi treasury would also receive its dues, which would support its ability to hold out in case the marketing of northern nationalized oil should falter.

According to this choice, the plan was drawn up and the articles of Law No. 69 were worded. It was not possible to state all the reasons behind this design. More definitely, it was not possible to say frankly that postponement of BPC treatment was a merely interim measure. In any case, there was no one at the Council of Ministers meeting who wanted to disclose this point, or to ask that the law should cover BPC in addition to IPC. The Council gave unanimous approval of the draft, and some must have realized the nature of the decision to postpone the BPC issue. Undoubtedly some could also see that nationalization of northern oil was enough of a hazard, but with a decisive command everyone dashed into the battle with a unified will.

But the foreign journalists did not delay, from the first moment, in trying to find out the real intentions about BPC. It was not really difficult for any observer at that time to fortell the intentions, though officials were not at all ready to disclose this card in the game of pressure. Even after the official settlement of the dispute with IPC,

Saddam Hussein used to answer, when asked about the future of relations with BPC:

> "It is no secret that we have no objection at all to dealing with any party or company on a commercial basis when it does not impinge upon our sovereignty and interests, and does not contradict our strategy on the patriotic and national level. Before June 1, the situation was different: the foreign companies used to form an element of threat to our sovereignty and future. But after June 1, the situation radically changed. The rudder has come into our hands. But we warn against any situation that may develop in the future that may conflict with our interest and national sovereignty. Such a conflict will not be solved in any other way but that which will be to the benefit of our sovereignty and national interests."

Now the plan was clear and the preparations complete. The decision was announced in the form of a law. Iraq had declared war, and the result of the heated conflict depended from this moment on its efficiency in conducting the battle. Any great plans and preparations would end in failure if the Command could not conduct the operations with courage and intelligence.

The Foreign Ministry summoned the heads of Arab diplomatic missions – one at a time – to inform them of the government resolution to nationalize IPC. The French ambassador was summoned for the same purpose.

Summoning the French ambassador with the heads of Arab diplomatic missions was a move which deserves special attention, though in fact it was not merely an expression of moral appreciation – it was an open message to the foreign interests.

France, starting with the de Gaulle era, had begun to move, in its foreign policy, according to a strategy which put French interests, to a large extent, in the first place, when those interests conflicted with the aims and exercises of American strategy. Applying this in the Middle East, France began to realize that there was no point in joining with the U.S. in a complete bias to Israel, and began to attempt to forge new links with the Arab world free from the old attitudes. It was natural that some of the progressive Arab states should respond to this new tendency, which could be useful to both sides, Arab and French. The French interests actually profited from this tendency, and French trade with the Arab world rose to high levels.

The size of this trade was 14.5 billion francs in 1969, but it jumped to 26.1 billion francs in 1973, which was an increase of 80%. In the field of oil in particular, France was foremost in realizing the changes in the oil world. In the field of relations with producing countries the victories of OPEC generally, the blows suffered by France in Algeria in 1971, the relative weakness of the French companies among the international cartel giants – all these considerations led the French policy to be more understanding in the conflict with IPC. It was natural that the Iraqi command should make use of the French rational attitude, even to encourage it by appealing to material interests.

It was this context that the French ambassador was received by the Iraqi Foreign Minister, and was informed of the RCC decision, issued with the nationalization law, as follows:

"In appreciation of the positive French attitudes towards the vital Arab issues, and as an expression of the wishes of the Iraqi Government to sustain and develop political and economic relations with France, and in order to protect the French interests in the oil of the operations nationalized by Law No. 69 of 1972, nationalizing IPC operations, the Iraqi Government declares its approval of entering into negotiations with the French side, if the latter expresses its desire thereof within a suitable period, in order to reach to a suitable agreement to insure the existence and continuation of those interests."

The development of events proved that this decision was a masterstroke.

And what about the Iraqi masses? The period of the first ultimatum ended early in the morning of June 1. And by 11 a.m., the period of the second ultimatum had ended too. Then day came to a close. True, the publicity machines and political organizations were saying that a statement by President Al-Bakr was to be braodcast, and true that the atmosphere was charged with excitement and all government systems and political staff were on the alert, but the day seemed to last forever, and it was natural that some anxiety should arise.

But by 10 p.m. all necessary measures were complete, and on the radio and TV, President Al-Bakr was declaring:

"In the name of the people and the Revolutionary Command Council:

"With reference to paragraph A of article 42 of the Interim Constitution, the Revolutionary Command Council decided, in the session held on 1 June, 1972, to issue the following Law:

"Article One: The nationalization of IPC operations in the areas limited to the company as per Law No. 80 of 1961. The state shall assume the ownership of all existing property and rights connected with the said operations, especially the installations and means of exploration, drilling, production of crude oil and gas, treatment, pumping, accumulation, transport, refining, storing, main and field-pipelines and other effects including the offices of the said company in Baghdad with all its installations and equipment.

"Article two:..."[4]

The revenge nurtured in the people's hearts for scores of years broke out. As soon as the President finished his broadcast of statement and law, the resounding masses began to crowd the streets in every town in the country.

NOTES

1. Saddam Hussein, *op. cit.*
2. *Negotiations and Nationalization.*
3. T. Aziz, conversation withe the author.
4. President's speech in no. 2 above.

* All information in this chapter comes from conversation between the author and Iraqi officials.

## CHAPTER 11

# LIBERATION OF KIRKUK

When Saddam Hussein telephoned to the Kirkuk Governor from his office at the National Assembly, he gave him a telegraphic message of two sentences: "Our wealth returned to us." and "A Command member is coming to you to assume responsibilities connected with the subject."[1]

Immediately, the Governor began to act, and moved the foreigners (only 20 in number) from the company headquarters as a protection measure – quietly and respectfully – to the first-class Kirkuk Hotel. They remained in the hotel until their departure. The new Iraqi administration offered them the chance to continue their work in the company in accordance with the nationalization law – but they all declined the offer. So, measures were taken for their departure.

For a long time, all communications made by the oil companies and the Baghdad office were under constant observation. Safety measures had been taken inside and around the company grounds in Kirkuk since the beginning of negotiations. Vigilance had been doubled since the RCC declaration of the two week ultimatum. Inside the company, every trustworthy person was contacted and given definite assignments. The Party staff, on various sites and shifts, kept a vigilant eye on things, and popular censorship doubled the safety measures. Security personnel were also put on the alert to face any attempt at sabotage. Censorship on telephones and external communications was stressed, also with regard to foreigners in offices and elsewhere.

Union headquarters had also issued orders prohibiting taking "anything" connected with the production process outside the company gates. This was to stop any attempt at upsetting production by smuggling some basic spare materials.

Saddam Hussein says that in addition to the general factors.

"There was a factor of a decisive nature: the daily general and detailed management of affairs connected with the nationalization decision. Some people may think that the political Command took the decision on 1 June, and left the management of affairs to the technical staff. What happened was quite different, and without belittling the importance and role of the technical staff, which made great and faithful efforts in preparing many details of the 1

> June decision, and the process which ended on 1 March, the management of the battle remained the responsibility of the political powers. There were comrades in the Command and the advanced staff who manipulated these affairs day by day. They did not only supervise the proposed plans, they were the ones who actually engineered them."[2]

We now come to the purchase of spare parts from abroad. This was previously done through the London office. There, all necessary data were available about the Kirkuk systems, equipment and requirements for importing them. There, all contracts were concluded and agreements made on manufacturing contracts to facilitate the arrival of goods, on time, at the stores in Kirkuk. Conditions were different now, and the Iraqi Company had to make the necessary changes in his organizational structure to fulfil this job. Making extensive imports demanded an expert staff and complete precision that was not available in Kirkuk. Therefore, a few hours after nationalization, a senior expert in foreign purchases was summoned from the State Company for Development and Planning of Oil Projects. The man was immediately charged with ordering the necessary spare parts and the formation of a foreign purchase staff.

On the morning of 2 June, an order was given to make a quick inventory of spare parts in the company stores. Naturally, the company's modern system enabled such a survey to be easily done by the computor. Therefore, a map of the store assets was before the company chairman in a very short time. A special committee studied the situation, which was only to find whether there were any bottle-necks. It was not necessary at that time to make import-lists according to the correct technical and economic requirements.

In order to imagine the situation, we have to remember that the different types of spare parts needed for the supply of company stores exceeded 50,000 in number. This is not to say that spare parts were 50,000 *units*, but 50,000 types of parts. These varied in nature, some needing renewal every two months, others every three months, every year, etc...

Under normal conditions, these matters were calculated precisely; since there was no point in storing up on what was not needed. Unnecessary storage ties up capital and leads to economic loss. But in June 1972, it was not possible for a political leadership at the top of the administration to evaluate matters according to economic considerations. In the first place, the basic question was to continue production at any cost. While the conflict with the companies was not yet settled, it was necessary to provide for the possibility of boycott from the manufacturers. Therefore, the decision on this point was to prepare for the worst, and that meant to have in store a provision of spare parts to last for two years.

In the light of these difficulties, measures were taken as follows:

Telegrams were sent to companies known for the production of tools and spare parts.

On 3 June, a decision was made to set up a purchase department. The company summoned staff from various ministerial systems in Baghdad to take charge and train new staff.

It should be noted here that the company had very great powers. The political Command issued orders to all government departments to respond immediately to all needs deemed necessary by the Iraqi company for continuation of work. Therefore, delegating the necessary number of staff for 3 to 5 months did not cause any problem. The order was carried out immediately.

Now came the problem of financing the necessary purchases. The company treasury at the time of nationalization contained a few thousand dinars only, not enough to pay even the salaries. Therefore, orders were given to remit large amounts of money to support the administration in its new responsibilities.

This emergency measure was not enough. After a telephone conversation with the Minister of Economics, a branch of the Rafidain Bank was set up inside the company area in Kirkuk, with direct connections with other Rafidain Bank branches outside Iraq. The Minister of Economics offered the company the privilege of direct import without import permits, and without foreign exchange limitations.

What remained unsure was the response of the exporting manufacturers to the contracts offered by the Iraqi company. Most of the supplies were of British origin and most of these companies answered the Iraqi telegrams obscurely or indefinitely. To insure against this, the following preparations were made:

With the decision to import all materials for two years, it was necessary to make priorities, emphasizing the more important, then coming down to the merely important articles.

If there was a certain technical system whose spare parts were difficult to import, for any reason, a contract was made to import a completely new system doing the same job. For instance, when a turbine used for pumping oil into the pipelines needed to import spare parts to secure continuation of its operation, and this proved to be difficult, another complete turbine was imported.

And, finally, the company resorted to the local manufacturing of parts, with the help of friendly countries, to provide substitutes for genuine parts.

With all these preparations, it was naturally profitable to use BPC which was not covered by nationalization at that time. Among the considerations taken by the Command in running the nationalization battle was, in the words of Saddam Hussein, "to raise a hammer threatening the companies if they were to attempt to confront the nationalization resolution".[2]
B P C was a pawn on which the hammer could come down at any moment. It was not necessary that the hammer should come down with the blow of nationalization. The threat could be limited, such as was the case with spare parts.

The Company summoned Mr Gillan (manager of IPC in Iraq) from the Kirkuk Hotel on the day following the nationalization. A meeting was held between the representative of the Iraqi state and an IPC representative in the manager's office. But, for the first time, the representative of the Iraqi state was sitting behind the desk, while the IPC representative was sitting on a chair on the other side. What a turn of the wheel!

In any case, conversation was about a few points that needed settling, such as departure arrangements for the foreign families. But during that meeting and conversation, the chairman of the Iraqi company happened to remark that supplies of the nationalized company were similar to those of BPC which was not nationalized, the same design and the same make. He added that the management of the nationalized company was ready to co-operate with BPC in the field of providing spare parts. There may be items at the nationalized stores that were lacking at BPC and vice versa.

Undoubtedly, Mr Gillan realized that the Iraqi official did not really mean co-ordination in the field of import. To safeguard against misunderstanding of the hint, the chairman of the Iraqi company said frankly: Isn't it unbelievable that we might have to stop production, transport or shipment because of a shortage of certain equipment, while the very things we need are imported by BPC into Iraqi territory, through our ports and airports? Can we let this happen and only stand by and watch? "

The message became very clear. And really, it was not possible to let the monopolies conspire to block production in the North and be allowed to continue their operations in the South with safety. The question did not necessarily call for state intereference. The "workmen" who prevented anything being taken out of the IPC territory before nationalization could also hide away some of the boxes that reached BPC, for instance.

Anyhow, the message was delivered, and the conversation continued. But it was clear that Mr Gillan immediately telegraphed the content of the warning he heard. Most of the manufacturing companies which had given ambiguous answers now sent their replies in a very clearly understood language.

After this, it was decided to send a team to Britain in July to make direct contact with manufacturers. The team was given full financial authority to sign and make contracts to import any items needed for the work as soon as the other party agreed. The team was sometimes met with some reservation in the first interview. But in the second one, the manufacturers usually approved the sales. Information available to government sources confirmed that representatives of the manufacturers used to report to the IPC headquarters in the period between the two meetings, and the company gave the manufacturers permission to export. The message had been received by IPC.

The assignment to making contracts for the import of spare parts was successful. But acting on the principle of expecting the worst led to some lack of confidence in concluding these contracts. Six months or a whole year sometimes passed between signing the contract and the actual delivery of badly-needed items in Iraq. Some interference seemed to be occurring in this period. Therefore, instructions were given to the team make visits to France, Holland and other countries after signing the contracts in London. These visits were to inspect the products of other manufacturers that were similar or alternative to the British, to safeguard against any emergency. No doubt those visits in themselves were a means of pressure on British manufacturers to undertake implementation of contracts with full precision.

All these measures were governed by revolutionary political perspective; but the political perspective did not ignore the economic aspect. Political action becomes inadequate if it leads in the end to an economic loss. Also, what is economically useful should not be assessed according to partial or temporary calculations in the first place. In this respect, the policy of importing spare parts on such a large scale could be considered a waste of income. But fortunately, with the wave of inflation which followed and exceeded all expectation, the operation was profitable despite the scale.

Thus continued the enthusiastic action, which achieved wonders. The association of political perspective with technical experience can solve any problem. The oil from the northern fields continued to flow without interruption. The Iraqi staff completely succeeded in controlling the management of production operations, so concern was directed to the development of administration and production. In the first field, the company systems were reviewed to be kept in line with the state laws. In the field of production the programmes laid prior to nationalization were revised and complemented, and were then put into operation.

So we come to the first well to be drilled after nationalization, the Well of Peace and Solidarity. The drilling was completed on August 3, 1972.

## NOTES

1. Information in this chapter comes from:
   A. Conversation with officials and technicians of the Ministry of Oil and Minerals and of the Iraqi Company for Oil Operations.
   B. Iraqi Company for Oil Operations, monthly report on oil operations, June 1972, first report after nationalization, unpublished.
2. Saddam Hussein, *op. cit.*

CHAPTER 12

# FIRST AND SECOND BREAKTHROUGH

On the day following nationalization, the *Financial Times* said that the Iraqi measure was "the most violent measure that has yet been taken by any oil-producing country... The general impression is that IPC share-holders preferred a full trial of strength and risking the danger of nationalization than to make a concession about compensations [Law No. 80]". Why were the share-holders so obstinate? The well-informed paper answers that "their excuses were probably based on the assumption that Iraq, which was highly dependent on oil revenues from IPC, could not bear a complete break-up. In the short and middle term, it was difficult to see how Iraq could market INOC production in any appreciable amounts, thought it was almost certain that the Iraqi Company for Oil Operations, which is owned by the state, could continue production".

These were the calculations – the Iraqi staff could continue the production operation, but the Iraqi Command would fail at export. These calculations were, in fact, not without basis. The Command also realized that the question of marketing was the greatest challenge. The Delegated Chairman of the Iraqi Company was right when he explained that the decisive element in the battle was the efficiency of the central political Command, and the success of marketing in particular. So what happened in the question of marketing? It began as a difficult problem; how did it end?

A clue to how it ended is contained in this scene at the settlement -negotiations between the Iraqi state and the delegation of the companies, whose operations in the North had already been nationalized. The delegation was insisting on getting some of the nationalized oil. When the Iraqi delegation asked why things were different since the companies used to say that Kirkuk oil was unwanted, as it had a weak competitive position, the answer was "We want it now. We want Kirkuk oil."

"In what quantities? "

"Between 20 and 25 million tons."

Here the three members of the Iraqi delegation could not help smiling. One official among them said: "We are sorry indeed. We wish we had the amount you need. You are not buying oil now by concession contract. You are asking to buy our oil on a commercial basis, just like the others. But, unfortunately, there are only 6 million tons of Kirkuk oil left now."

At which Danner, the chief company delegate, said: "Please repeat the figure once more! " The Iraqi official answered: "Of Kirkuk oil, 48 million tons were sold. There are only 6 millions tons left".

Danner was an expert, and an experienced negotiator. But he could not hide his shock and frustration. He was stunned into silence, and he began to fumble with his long whiskers, with agitation that was apparent to the Iraqi negotiators. After a few moments of silence, Danner said again: "This is unbelievable. There must be some mistake. Please, is what you say true? "

"The figures given are real figures. Why should we hide the oil inside the ground or produce less than our top capacity? It is to our benefit to sell, in order to secure our revenue."

This, then, is how the marketing problem ended. The nationalized oil had not been blocked, and lost was the main tool of pressure in the hands of the companies. He who laughs last laughs best! But in our case, the laughter came after a gigantic effort.

The story began in fact when INOC prepared to undertake marketing. The company found itself in the midst of difficulties. The simplest question was a riddle. How were the crude oil contracts worded? What were the documents needed for shipment? How were loading programmes prepared and arranged? Who were the real buyers of crude oil in the direct markets outside the international cartel limits? What were the conventions and rules followed in payment of crude oil prices? etc.

In spite of all this, the company, backed by the revolutionary willpower, had to plunge into direct development, and had to market the produced oil. Therefore, INOC contacted a number of independent companies and agents and friendly consuming governments in order to conclude contracts to sell crude oil, in spite of BPC threats to prevent those activities. And the companies actually carried out their threats. On April 6, 1972, a little before Iraq was planning to start to market the Rumaila oil, a press statement was issued from BPC in the form of a warning to Iraq and other parties concerned in case they thought of dealing with the Rumaila oil. The statement said that the oil was extracted from the company concession area, and therefore was the property of the company, and asked foreign companies not to conclude contracts to buy the stolen oil. In accordance with the warning, a court action was brought against INOC in Ceylon, a tanker was held at Capetown, and another action was brought in Italy.

If attempts were so feverish in trying to blockade the Rumaila operation, what would it be like trying to market the oil from the main fields? The challenge was great, especially with lack of experience. In learning how to extricate themselves from this they learned the skill of sound strategic visualization of international oil relations. It was necessary to concentrate first on markets in the Socialist and non-aligned countries. The preparations for the nationalization battle had included sending representatives from President Al-Bakr to Socialist countries who signed a number of agreements, between April 7 and a little before June 1972, to barter for Iraqi oil. The agreements also secured loans to be paid up when possible. It was announced at the time that barter agreements were based on Rumaila oil. But, in fact the contracts largely exceeded the capacity of Rumaila field at the time (5 million tons/year). Also, Iraq had been careful not to specify in the contracts that the bartered oil was Rumaila oil in particular. The text of the contracts said that barter was to be done for Iraqi oil, hence there was no legal objection – after nationalization – to dealing in Rumaila or Kirkuk oil. Naturally, friendly relations with these states did not allow for interpretations contrary to the text.

So, by virtue of these agreements, there was the security of a minimun level of marketing for the Kirkuk nationalized oil. Links made with other markets before nationalization, for Rumaila oil, were also good experience. In these markets, barter agreements were offered, and the principle of deferred price was accepted by Iraq, to facilitate and encourage trade, but the principle of reduced price was not accepted. The Iraqi Command was careful, even in the severest situations, to protect the price which was based on a structure decided by OPEC, and had previously refused an offer from the companies threatening this structure in return for raising production levels.

Because of this limited experience in probing the markets, the Follow-up Committee had to prepare for the export of about 60 million tons of national oil from the North and the South. It was understood not to be possible to export all that amount to the Socialist countries. The proportion of oil in the energy consumption structures of European Socialist states was still limited. The Soviet Union could be excluded from this generalization to a certain extent, since it owned local resources of crude oil, and also exported oil to cover the greater part of the needs of neighbouring Socialist countries. Soviet exports to

Poland, the German Democratic Republic, Czechoslovakia, Hungary and Bulgaria formed 98.7% of the total oil imports of those countries in 1965. The figure became 98.4% in 1966, 95.1% in 1967, 97.7% in 1968, 95.8% in 1969, 95.2% in 1970.[1]

Therefore, the estimates from the start were that the markets of those countries could not import much of the large Iraqi production. True it was expected that Socialist countries put strategic and political considerations first, and should arrange to buy quantities of nationalized Iraqi oil despite pure economic calculations. But the situation, for Iraq, remained merely the need to secure a minimum level of marketing. This does not belittle the value of approaching this market, since securing a minimum level at that stage was considered an important breakthrough of blockading attempts.

But how were things on the western front, and with the cartel?

The Law of Nationalization was announced on the evening of 1 June 1972, and on 2 July, the British Foreign Ministry issued a statement saying: "The British Government regrets that neither the Iraqi nor the Syrian Government informed Britain before taking nationalization measures against IPC. The British Government, which supervises IPC, is now making contacts with the other governments concerned". What is noticeable here is the arrogant accent that Iraq had to ask permission before taking a decision that lay within her sovereignty limits.

On the same day of 2 June, John King, the spokesman for the US State Department, stated that his government hoped that Iraq would make immediate and suffcient compensation for IPC possessions. What should be noted here is the emphasis on immediate compensation which is the usual way of opposing a revolutionary law of this type.

On 3 June, an IPC spokesman declared that the company intended to take legal measures against any buyer of oil coming from nationalized wells whenever that was possible. The spokesman charged the Iraqi Government with responsibility for the conflict, since the Government did not want to come to an agreement, and showed no sign of readiness to negotiate. He said that the Iraqis seemed to want to accelerate the problem for a reason still unknown.

But this unknown reason was "uncovered" by BBC on the same day, quoting *The Guardian:* "By nationalizing IPC, the Ba'thists may have put an end to four years of sarcastic Arab remarks that the Ba'thists are people who talk but do not act. Yet, this action may, when things run contrary to their desire, lead to their downfall". The BBC meant that the question was merely a foolish move for show, and would lead to disaster.

In fact, I have copied this comment by the BBC from one of the reports presented to the Follow-up Committee. I find no objection to quoting the entire report, since it reflects one aspect of the inside picture.

To: The Chairman, the Follow-up Committee
Dear Sir,

Please be advised of a summary of today's news' reports concerning the nationalization situation:

1. Raymond Edde [a former Lebanese Minister] stated that marketing Iraqi oil will be faced with the price Iraq demands. He added that Saudi and Libyan oil is cheaper than Iraqi oil as it is shipped to Europe by huge tankers, while Iraqi oil passes through pipelines which entails various charges on the producer. He also added that the company reduction of production is caused by the high cost. Edde asks: "Can Iraq reduce the price of oil to European markets? "

2. Beirut News Agency said that the Algerian Government declined to attend the OPEC conference to be held in Beirut.

3. The Lebanese Minister of State for Industry and Oil Affairs denied that his country was to nationalize IPC installations, "since that runs counter to a free economic system". Lebanese desire to support Iraq may lead to their buying the pipelines on Lebanese territory from IPC, if the company would agree to such a purchase; or Lebanon will have to build a new pipeline to facilitate and secure the passage of oil.

4. French quarters denied reports that France agreed to hold a meeting in London for representatives of states affected by the nationalization. France finds it useful to continue contacts with oil companies to find out points of view. Responsible quarters say that planning such a meeting is premature, because there is not enough information about the various aspects of this highly complex situation.

5. The BBC quoted *The Guardian* (as above).

6. The Iraqi dinar gained 35 fils in Abu-Dhabi.

7. The British Foreign Minister stated that consultation will be held among Britain, France, Holland, and the U.S. next Thursday on problems stemming from nationalization. He said that he had had a preliminary exchange of opinions with his French and

American counterparts about the subject.

He confirmed that his government expected that Iraq and Syria should pay real and suitable compensation soon. He called on the Governments of Iraq and Syria to secure the safety of company employees. He added that the British Government regrets not being previously informed of the subject.

8. Amman radio said that Saudi Arabia asked to postpone the OAPEC meeting to the eleventh of this month: Saudi Arabia sees that this meeting should be held after, and not before, the OPEC emergency meeting.

9. The Eighth Arab Oil Conference concluded its sessions in Algeria and the resolutions will be submitted to the Economic Council of the Arab League.

In this concentrated report, Saddam Hussein commented on two points, one of which was the second point about the Algerian attitude, near which he wrote: "1. Verify the news by contact with the Foreign Ministry. 2. Find out the reasons".

The other point that called for a comment was No. 4, about France, and his comment was: "Ask Dr Sadoon Hammadi to provide us with what he has about this point, and about the issue in general".

The report was one of scores of reports submitted that day. The instructions I quote here are a drop in the sea, yet they are an indication of the Command's awareness of enemy manoeuvres. Within the axis of the oil-producing countries, there was a special sensitivity about the Algerian move. For the western countries, there was an alert follow up development in the French attitude.

We shall now follow these two pointers, beginning with the French attitude.

From the very beginning, the plan was to encourage France to adopt a more moderate attitude by giving it a most favoured treatment. It is notable that the RCC decision excluded *France* from the economic results of nationalization, and not the *French Oil Company*. Nationalization was applied 100% in IPC shares, and the French Oil Company was not excluded from the RCC decision, provided that the Iraqi Government was ready to enter into negotiations to reach a suitable version securing the French interests. When the decision provided that, it was meant to exclude France from the negative economic results of the nationalization decision, and to offer the French Government a chance to make contracts to import its previous share of Iraqi oil, at the same concessionary prices, and for a period of ten years. Naturally the contract would secure material interests for France since it would continue to receive the same gains and revenues which it used to get before nationalization. But these would now come by means of a commercial contract, accepted by the Iraqi Government, as an expression of French desire to support special Iraqi (or Arab) relations with France. The RCC decision was an occasion for concentrated meetings and talks. When an Iraqi delegation went to hold reconnaissance talks, a few days after nationalization, with British and IPC officials in London, the delegation stopped in France first, and also had meetings in Italy.

According to the French papers of that period, various pressures were being exerted on the French Government, and it was not easy to reach a decision. But on 7 June, "the fish began to bite" when the French Ambassador delivered to the Iraqi Foreign Minister a message saying that the French Government did not object to nationalization, since it was compatible with international principles of the sovereignty of the state. The message also said that the French Cabinet studied with special interest the offer presented by the RCC to hold talks in order to reach a plan securing French oil interests. It is now known that after this message Saddam Hussein agreed to go to Paris to head an Iraqi delegation to the talks. Therefore Baghdad received numerous representatives of the French Oil Company as an official representative of the French Government, to make preliminary contacts. Yet the question remained undecided. *Le Monde* said on 9 June that the French Government delayed a little before accepting the offer of the Iraqi Government, and did not give a final answer until that date. Acting against the will of the European countries was a threat to European economic unity, especially because Britain, who had recently joined the Common Market, was the worst hit by nationalization. But *Le Monde* added that, on the other hand, there were those who said that France would accept the Iraqi offer, encouraged by the other western countries, so that the Soviet Union would not stand alone in the development field in the Arab world.

In fact *Le Monde* was giving advice to the French Government to accept the Iraqi offer, since France in 1970 was the third largest exporter to Iraq, and the second largest importer of Iraqi oil.

I think that the *Nouvel Observateur* article of June 12 painted another important picture of the various factors behind French hesitations and moves. The French company, according to the article, could not let down its partners at IPC. Those partners had previously supported the French company when Iraq wanted, during the Algerian

War, to nationalize the French share. In 1971 the international cartel supported the French company in the embargo enforced on the Algerian desert oil in order to get valid compensation. Also, the French Oil Company had shares with other IPC partners in a number of Middle East countries: Iran, Abu-Dhabi, Qatar, Oman, Dubai, to name a few.

But the French company tried to reach a settlement without harming its partners so that it could profit from the Iraqi offer of most favoured treatment, which was the result of French policy in the Middle East, after the company had succeeded in strengthening her situation in the Arab world, by virtue of the co-operation carefully nurtured with Algeria, since the settling of the question of desert oil nationalization.

An expression of the French company's hesitant attitude to upset the balance of conflicting interests was its constant contacts with the IPC London office, while at the same time sending delegates to Baghdad. The London IPC officials said to the French company that they hoped the French would openly reject the most favoured treatment offered them by Baghdad. But the Iraqi offer was difficult to reject so easily. The partners in London felt that the French interests were tending towards an independent attitude. The British Foreign Minister called for a meeting to be held in London on June 8 for directors of economic affairs in the foreign ministries of countries represented in the IPC (i.e. France, England, Holland and the US). But France declined to attend, saying that there was no exact information about the Iraqi attitude or the attitudes of other IPC partners yet. The French Government declared, in the words of the official spokesman, that France would attend the meeting to be held in Paris, on June 18, with oil experts representing the countries of the four partners in the nationalized company, within the framework of the European Development and Co-operation Organization.

The *Nouvel Observateur* concluded the above-mentioned article by saying that this confrontation "put the issue before M. Georges Pompidou, who had to maintain a difficult balance between the requirements of his Arab policy and solidarity with oil groups in London, New York, and The Hague, during his talks with Saddam Hussein".

Indeed Pompidou had to decide the question one way or the other. It was not conceivable that the French attitude would be completely swayed to either side, since France had interests with both parties. Yet, some bias here or there was inevitable.

On 14 June, Saddam Hussein was beginning his Paris visit. Pompidou had made his calculations and decided that his interests lay in accepting the Iraqi offer of most favoured treatment. The joint statement issued at the end of the visit said that the Iraqi Government decided to sell 23.75% of Kirkuk oil to the French Oil Company (CFP) for ten years, and according to financial and economic terms prevalent prior to the nationalization law. An agreement was actually signed, on the last day of the visit, between the Iraqi and French Governments, in accordance with this decision. On February 3, 1973 INOC signed a long term contract to provide the French company with crude oil produced from fields covered by the nationalization law, for ten years.

The trip by the Iraqi delegation to Paris in June, headed by Saddam Hussein, was a culmination of the Iraqi efforts to cause a split in the monopolies' front. The companies were hesitant from the first day out of fear for BPC. Then they despaired more of forming a unified front when the French Government preferred to act independently. The Paris trip and its results confirmed that the Iraqi Command had a firm grip on the initiative.

Yet there is a testimony recorded by the Iraqi officials. The June 18 agreement was largely achieved by the personal efforts of the late French President, Georges Pompidou.[2] When the file on that case came to his office, the French Premier was not enthusiastic, and there were influential quarters that supported his attitude. But after the first meeting with Pompidou, the green light was given to the French administration to sign the agreement. To decide the situation finally, Pompidou held another emergency meeting to finish the business.

So ended the second large breakthrough of blocking attempts. This time the breakthrough came inside the camp of the forces that were supposed to unite to enforce the blockade.

**NOTES**

1. Q.A. Al-Abbas, *Oil and Development Review* I,4.
2. T. Aziz, conversation with the author.

* Official Iraqi documents.

# CHAPTER 13

# ARAB SURPRISE

An emergency meeting of OPEC was held in Beirut on 9-10 June. It was immediately followed by an emergency meeting of OAPEC, also in Beirut, which adopted a resolution to support the Iraqi attitude, and recommended the call for other meetings, depending on the development of the situation, to take suitable measures, in collaboration with Iraq and OPEC. The conference also recommended a meeting of the Ministers of Finance of the member states in OPEC, to be held in Baghdad, to estimate the size and direction of financial support necessary. The meeting was actually held on 23 June.

Apart from the official statement, which was a form of moral support, important in itself, Iraq wanted to test the result of the Iraqi demands expressed in the messages of President Al-Bakr to heads of states before nationalization. In particular, a meeting with the oil-producing countries aimed to find out three things: Firstly, whether there was a chance of loans or financial assistance when needed; Secondly, whether there was a definite commitment that the oil-producing countries would not be lenient in raising their oil export rates to compensate the world markets for the nationalized Kirkuk oil; and thirdly the moves needed to prevent monopolies from blockading nationalized oil and setting legal impediments against its marketing attempts.

Concerning the first point, the OPEC meetings left the matter open for bilateral contacts. Libya and Kuwait promised to offer a loan of $100 million (although the Iraqi officials say that what reached the Iraqi treasury was really less than one tenth of that amount).

As for the second point, it was difficult to follow up the situation with complete precision, since these states themselves did not control the marketing of their own oil. Even with honest intentions, they did not have sufficient information stored up in the computers of the marketing systems, and they could not trace internal relations and agreements among distribution networks. But, within possible limits, the Follow-up Committee kept an eye on them, and made others aware of this activity.

Here is one example:

**Subject: Production Increase in Arab Gulf States**

1. Production in May 1972 18,1224 million barrel/day
   " in June 1972 15,3974 " "
   (the difference) -17.7%
   " in July 1972 15,8159 " "
   (the difference) + 2.7% from June)
   in August 1972 17,6781 million barrel/day
   (the difference +11.8% from July)

2. Main countries where increased production occurred:
   Saudi Arabia, Kuwait, Abu-Dhabi:

| | *July 1972* | *August 1972* | *Percentage increase* |
|---|---|---|---|
| A. Saudi Arabia | 5,4283 mbd | 6,2386 mbd | +14.9% |
| B. Kuwait | 2,8188 mbd | 3,5989 mbd | +27.7% |
| C. Abu-Dhabi | 0,9666 mbd | 1,1396 mbd | +17.9% |
| D. Iran | 4,9518 mbd | 5,0034 mbd | + 1%. |

3. Percentage increase between August 1971 and August 1972

| | |
|---|---|
| A. Saudi Arabia | +32.5% |
| B. Kuwait | +24 % |
| C. Abu-Dhabi | + 25.2% |
| D. Iran | + 6.9% |

4. Evaluation of Reasons for Production Increase:
   A. Economic boom in 1972 led to consumption increase, then, to increase in production.
   B. Japan increased imports after ending port labourers' strike.
   C. Kuwait had previously called for freezing of production to preserve oil wealth. So, the increase in August denotes a change in oil production policy in Kuwait.

5. Conclusion:
   A. It is necessary to find out the reasons for these exceptional increased figures and the relationship between them and the possibilities of compensation for nationalized Iraqi oil. Therefore investigation must be made to find out the markets to which this oil has gone, and whether these amounts have gone to the traditional markets for Iraqi oil.

   B. It is imperative that Iraqi diplomatic establishments abroad should participate in these investigations. The Arab Gulf states themselves could be asked through Foreign Ministry channels...

Awaiting your advice.

I am not familiar with the results of the investigations suggested by this report. But it seems that the Iraqi officials felt that the Arab oil-producing states did not address themselves to this question with the necessary concern.

There remains the third point. Some contacts were made with the companies to stop blockading attempts. The Secretary-General of OPEC, at that time, was asked to take part in mediation so that the companies would not resort to a blockade, and at the same time, to prepare the way to decide upon the impending problems.

Among the Arab states, there are two who played a special role in the issue of Kirkuk oil, because of their geographic situation. These were Syria and Lebanon.

The path of this oil to the Mediterranean shores had to pass through pipelines lying in Syrian territory, some ending at Banias terminal, others continuing through Lebanese territory to pour into Tripoli terminal. Because of this, the co-operation of these two countries, especially Syria, in the nationalization battle, was a question of life and death for the entire battle. Otherwise what would be the use of conducting the battle at all? What would be the use of continuing production and oil pumping successfully, of destroying the cartel blockade on marketing and concluding export contracts if the oil could not reach the tankers on the Mediterranean shores?

It was expected that the Iraqi Command should resort to some reduction in price to help export movement. Such reduction in price would only be possible if costs could be reduced. If the Iraqi-Syrian relations at that time were of the nature of a unified struggle, it would have been conceivable that the Syrian authorities would reduce the transit rates below the figure received from IPC until nationalization crossed the difficult period. On a purely economic basis, if the success of nationalization were an objective agreed upon, it would have been in the interest of the Syrian economy to make such temporary reduction in transit dues since it would have reduced export costs and increased exports, and ultimately, the increase of income would be an added benefit to the Syrian treasury.

Anyhow, there was no such optimism on the Iraqi side concerning the Syrian attitude. The real surprise was that after nationalization of the pipelines, Syria doubled the transit dues. These dues were 22 cents per barrel, but the Syrian Government decided to raise that to 45.55 cents.

The Iraqi Command had expected that the nationalization profit in the first year, 1972, would exceed 18 cents per barrel. The nationalization profit was the difference between government revenue from selling one barrel after nationalization and similar revenue in the time of the concessionary companies. These economic calculations were made according to market indexes in that year. If the nationalization profit after these calculations was 18 cents per barrel, then the increase in Syrian dues would swallow more than the nationalization profit. This meant that in the case of successfully marketing an amount of oil equal to a traditional amount before nationalization, the revenue of the Iraqi Government, after all this struggle, would decrease.

Yet, there was nothing to do but give in. It was essential to suppress news of the dispute. When the Iraqi Minister of Oil and Minerals visited Syria and Lebanon immediately after nationalization on 3 June he stated: "Agreement was achieved on all questions connected with the new situation stemming from nationalization". He said that he was "pleased with the results fo the visit in which we were in agreement on all the questions discussed".

It was necessary to give in without a cry of pain. On June 8, Iraq was signing a temporary agreement with the Syrian Government... in accordance with the Syrian terms. Talks were resumed which lead to the 18 January 1973 agreement.[1] Iraq was obliged again to accept the terms in order not to face the negotiations of the companies while the arrival of the oil at the sea was not ensured.

The agreement period was three years. The transit dues were fixed at 45.55 cents per barrel, and the price of Iraqi oil used by Syria locally was fixed at 2.55 cents per barrel in 1973, 2.65 cents per barrel in 1974, and 2.75 cents per barrel in 1975.

As for Lebanon, it was also included in the "complete agreement of opinions" mentioned in the statement of the Iraqi Minister. The fact is that the Lebanese Government succumbed to pressures by IPC, and months of meetings failed to change the Lebanese attitude. The Lebanese Government refused to nationalize the pipeline "as this contradicts the principles of a free economy! " The Lebanese Government also refused to negotiate with IPC to buy or hire the pipeline, which had stopped piping nationalized oil to Tripoli. When Iraq concluded the March agreement which settled the problems with the companies, it included an article that could solve the problem with

Lebanon. IPC agreed to return to the Iraqi Government the possession of immovable property, comprising transit installations and the terminal in Lebanon, exclusive of the Tripoli refinery, storage tanks, loading and unloading facilities. The company provided for the validity of these terms subject to written consent from the Lebanese government before 31 December 1973. Naturally the Lebanese Government did not agree, but furiously announced occupation of pipelines and installations in 48 hours. So, an agreement was signed between Iraq and Lebanon (after the agreement with the companies) on 5 March 1973. Lebanon insisted on having the same treatment given to Syria, and it was given that. Piping of oil to Tripoli was thus resumed.

## NOTES

1. Text of the Syrian Agreement in, M. Al-Mousawi, *op. cit.*

## CHAPTER 14

# AFTER THE WHITE FLAG

The operations in Kirkuk had to go on. Pumping the oil could not stop. The masses all over the Arab world were waiting to see what would happen. If they stopped, memories would be stirred of the nightmare of the Iranian failure. There, before the collapse of Musaddaq and his experiment, production had stopped and labourers had been laid off.

The continuation of production and pumping were essential to keep morale high, and to confirm confidence in the efficiency of the national staff, and in the success of nationalization. Systematic production needed the passage of oil outside Iraqi frontiers, and this was ensured by the signing of an agreement with the Syrian Government, though at a high price. But what next? What about the tankers to be filled with oil from Banias storage tanks? What if those tankers were delayed by marketing problems? .

The outlets of nationalized oil were restricted in this case to Dora and Hadeetha refineries in the Iraqi territory. In addition to this, Homs refinery (owned by the Syrian Government) had to be supplied with crude oil, as did the Tripoli refinery which was directed by IPC in Lebanon, though its products were distributed and sold inside Lebanon in accordance with a special agreement between IPC and Lebanon.

What about the rest of the oil? It was directed to the storage-tank farms . But with the complete stoppage of export during the month of June, instructions were given to reduce rates of pumping from the fields; when the storage-tanks were full, pumping came down to the lowest levels. Saddam Hussein's decision was explicit, however: "Do not stop production under any circumstances"[1] And work in fact did not stop for even one day in Kirkuk.

At the same time efforts continued to be made to find markets. Export to the Socialist countries began, and the agreement with France was another important step. But the monopolies did not stop the counter attacks. They were still insisting on refusing the right of Iraq to market its oil. In the various meetings with the companies, they tried to use compensation as a means to penetrate to a position similar to their former one, hence the decisive words of President Al-Bakr on 17 July 1972.

"We now declare that we shall never allow any attempt to make the question of compensation a means by which the companies try to

> penetrate and side-step the nationalization law, or impede its implementation. The companies, which should know by experience that we mean what we say, should realize this fact and proceed according to reason and logic."

Before these words by the President, there was the trip to the area by William Rogers, then the US Secretary of State. The Iraqi authorities knew that his trip implied an attack on the Iraqi nationalization and that he tried to enlist the enmity of the countries he visited against the Iraqi policy. On 13 July, the Iraqi Foreign Minister stated to *The Times* that "all British interests in Iraqi will be liable to liquidation [naturally this was an allusion to BPC] if the British oil companies try to impede the Iraqi efforts to market the nationalized oil".

But IPC paid no heed. So, on 24 July, the Ministry of Oil and Minerals made a statement saying: "The said company lately started a publicity campaign, through warning letters and press advertisements, claiming rights to the crude oil produced and exported by the Iraqi Company for Oil Operations". The Ministry explained that this attitude was contrary to Iraqi laws, international conventions, U.N. resolutions, international courts' jurisprudence and national courts all over the world.

> "Believing in the justice and legitimacy of our attitude, and in the invalidity of the IPC claims, the Iraqi company confirms readiness to take all necessary measures, according to agreement, to ensure for all buyers, or those willing to buy the Iraqi oil, the full enjoyment of rights and facilities arranged for them by contracts of buying crude oil from the Iraqi establishments concerned."

Export of nationalized Kirkuk oil began in July. The Iraqi Command informed the companies that any interference with that oil in the high seas would be answered by the nationalization of BPC. The amount exported in July was 583.6 thousand tons. In August the figures jumped to 1.7 million tons. In September it jumped again to become 2.3 million tons.

The blockade was cracking and success was accelerating. From this success, it was possible for OPEC mediations and French interventions to play a positive role between Iraq and the companies. An IPC spokesman declared that mediation had succeeded between the company and the Iraqi Government. Therefore IPC would postpone taking legal measures towards oil-buyers until December 1972.

On 25 July, the Yugoslav paper *Borba* published a statement by an

Iraqi Command official. The statement declared that with reference to achievements, calculations indicate that the most important problems of marketing nationalized oil would be solved within 4 to 6 months. Those calculations were proved to be true. The average of export figures in September approached the top export capacity of Banias terminal. In October the Banias exports scored more than 3 million tons. In December a new record was reached: 3.7 million tons.

It is true that export was completely stopped from Tripoli until March 1973. True also that the total amount of Iraqi oil exported had come down, and with it the state revenue. But who said that nationalization was supposed to pass without difficulties or sacrifices? It was a real success that Iraq did not have to face the case of "the worst possibilities", which was expected. By virtue of the clever administration of the battle, the sacrifice was at a minimum – BPC did not dare stop production for one day. When export of Kirkuk oil was stopped in June, BPC was exporting 3.2 million tons, and kept this rate up in the following three months, (which meant that the Iraqi plan with BPC achieved its precise aims, and the Iraqi Government received full dues). The total of Iraqi oil exports reached about 55.9 million tons between June 1972 and March 1973. In the same period a year before (June 1971-March 1972) the figure was about 65.2 million tons, a decrease of 10 million tons; a small tax indeed! Yet, the figures of the later months of the period formed a clear indication that the blockade against Kirkuk oil was beginning to crack. The companies realized that something of the sort was happening, but they did not realize the extent of the crack. Nor did they realize that contracts included, and brought to light by, the Iraqi delegation during negotiations had finally decided and terminated the problems of marketing.

The Report of the Eighth Congress shows that reaching a unified plan for a solution of outstanding problems with the companies was difficult, and a cause of some controversy. The report says:

> "During that stage, we were faced with two different viewpoints: one greatly exaggerating the political aspect in the nationalization operation, demanding we show no flexibility whatever in the question of compensations, and the other biased towards the economic aspect. This was prompted by the difficult economic situation coming out of nationalization, and by movements and criticisms or reactionary forces and defeatists. These preferred the economic aspect, saw the necessity of coming to an agreement with

the companies, and showing flexibility in various matters, in order to resume a natural economic and financial position to bypass the critical stage.

"Both viewpoints were wrong in handling the issue, when adopted separately. The first starts from a formal revolutionary position, not taking into consideration that a revolutionary step must be accompanied with a degree of flexibility and tactics, which helps towards success without affecting its essential features, and that the nationalization process, despite its serious political aspect, must be calculated in the light of happiness it can secure for the people in real life. The second opinion stemmed from the fear of the consequences of a serious operation like nationalization, and did not have the imagination that presupposed a high degree of fortitude and confidence in the Party and the people in the management of a serious battle of this sort".

The Report adds that the Command prepared a precise analysis on the subject, and laid down definite principles and a political basis for the management of negotiations, emphasizing the necessity of flexibility in the questions of a legal nature.

"The Command started from two basic points in their analysis. The first was to preserve for nationalization its implications and political dimensions on local, Arab, and international levels. The second was closely connected with the first point, implying that the nationalization battle has its financial and economic aspects, and that it must be conducted and concluded on the basis of securing great financial income for the country, and making the people feel, after winning the battle, that nationalization was of great economic importance for the people, in addition to its political, patriotic, national and revolutionary implications.

"After the Command had decided these sound bases, the negotiations were conducted accordingly. The companies, whether through representatives who undertook to negotiate at first, or in the direct meetings later on, tried to sound our ability to hold out and conduct the battle.

"The companies found out that the Revolution was firmly decided not to retreat from its decision with all its basic implications. The companies' intelligent skills and tactics were faced with decisive attitudes and more intelligent tactics. The Command supervised the battle directly, not leaving it to officials and technicians, in spite of their activities and faithful efforts, handling every point, large or small, in a politico-economic perspective, interconnected and complementary."

In the end, the companies gave in and signed the general agreement in Baghdad on 28 Februray 1973. At 1.25 p.m., 1 March, President Al-Bakr was making the statement of the complete victory.

**General Agreement**

"1. The Iraqi Government on one side, IPC, BPC, MPC, with shareholders and associated companies on the other side, in accordance with laws and regulations valid in Iraq, especially Law No. 80 of 1961, amended by law no. 24 of 1970, Law No. 125 of 1967 with its amendments, Law No. 97 of 1967, Law No. 229 of 1970, and Law No. 69 of 1972, agree that the reciprocal undertaking affixed hereunder is a final settlement of all pending questions and claims owed by either party to the other.

"2. The companies undertake to pay, etc...[2]".

This long series of laws and amendments formed milestones on the road of the liberation of oil wealth by legislation. The companies were no longer able to practice a merely formal recognition of these laws, since the attempts at blockading had failed, and a completely new stage had begun with progressive national revolution, leading to a new stage in international oil policy.

The March agreement cleared the old accounts to start a new page. Like any acocunt, it had to be expressed by figures. It should be mentioned that on the day following the nationalization[3] the companies estimated the real value of assets covered by the nationalization decision, and consequently to be covered by compensation, at a sum between 520 and 780 million dollars. With such an exaggerated figure in mind, the companies started applying pressures and seeking bargains. Added to this, of course, were the claims of the companies for compensation for the results of Law No. 80, which was represented in buying 4 billion barrels of crude oil (550 million tons) produced by INOC at a reduced price (162 cents per barrel). At that time, this meant that Iraq would have lost 300 million dinars (based on the difference between the sale price demanded by the companies and the lowest price Iraq was supposed to get on the market). In addition to this sum of 300 million dinars, there was another claim by the companies, to complement the claimed compensation, that was the 7% freely demanded of INOC. Measured by production of that time, this meant another 200 million dinars, thus bringing the sum total to 500 million dinars of loss. As for what Iraq was due from the companies, in the last negotiations, this was very "generously" estimated by the companies not to exceed 100 million dinars. The final total would thus have the companies claiming 400 million dinars (1352 million dollars) from Iraq.

On the other hand, Iraq estimated standing claims against the companies, in the negotiations prior to nationalization, at about 600 million dinars. Even when this estimate was accepted, and with compensation for supplies according to previous estimates by the companies, Iraq would be indebted in the final settlement.

But, with the success of nationalization as a basis – that is, with the change of balance of power in the months following nationalization, and with the tactical success in manoeuvering – the final figures in the March agreement specified that the companies should pay to the Iraqi Government a total of 141 million pounds sterling. In return, the Iraqi Government would deliver, or ensure the delivery of, 15 million tons of Kirkuk crude oil, an amount equal to about 315 million dollars ($3 x 7 barrels x 15 million tons) that is about £128 million. This sum covered IPC immovables in Lebanon which the company agreed to sell to the Iraqi Government.

In addition to that, the general agreement provided that the companies undertake to do their best to speed up development programmes in the BPC area, in order to raise production capacity from 35 million tons in 1973 to 80 million tons in 1976.

It must be mentioned here that the basis of accounts settlements was discussed, within the general framework of negotiations, by the Council of Ministers, after the approval of the basis and framework by the Party Command.

Now, what about the Kurdish Democratic Party, whose command at that time was in the hands of Mulla Mustafa and his faction?

The March Declaration (which recognized the national rights of the Kurds) was instrumental in preparing the Iraqi masses to move together towards the aims of the democratic, national revolution. It was expected that progress within the context of this objective could falter – as did the other Front endeavours – under the influence of inherited and reciprocal bias. In the case of the Kurdish issue, there was, in addition to these biases, a planned separatist endeavour led by the Barzani movement, and an official programme intended to sabotage the efforts towards the achievement of national unity, in order to preserve the backward feudal relations and to conspire with outside forces. The

Barzani movement was, accordingly, on the side of all the forces which were conspiring against the revolution in Iraq. The threat of a Barzani unity in the North, to deplete the revolutionary resources of the state, was one of the weapons being saved for an opportune time.

It has now been proved – from US Congress investigations of CIA activities – that Barzani was actually moving towards this in those days, and that he made contacts with the CIA representative in the Middle East, in his headquarters in Tehran. Those investigations proved that the US official made a recommendation to his Government to give the needed support, and was backed by the Shah in a later communication. Although these data are confirmed by documents of those days, the possibility of a Barzani conspiracy had been calculated at that time. The Revolutionary Command saw to the prevention of that possibility, and to the postponement of its occurrence for the longest period possible. There is no doubt that the March Declaration, and the political efforts that followed, were a successful step in that direction.

With the acceleration of conflict with the companies, President Al-Bakr explained, in the meetings of the Council of Ministers, the complications of the Kurdish situation, and the probable Iraqi response, without disclosing the cards the Command had up its sleeve. (There were, in the Council, three ministers representing the Kurdish Democratic Party.) That very acceleration of conflict with the companies was among the factors which paralysed the possibility of open revolt by the Mulla against the revolutionary rule.

When those accelerating events culminated in nationalization, the Mulla and his faction were taken by surprise. In the midst of general overwhelming enthusiasm which, naturally, included the Kurdish masses, the Mulla had no choice but to pretend to support the nationalization decision. But in his private meetings, he expressed discontent with the position it put him in, which paralysed his power to move.

Despite this, the possibility of a move by the Mulla had been calculated. General Hammad Shihab, the Minister of Defence, stated immediately after the declaration of nationalization that the decision heralded a new revolution, and that the Iraqi army was fully prepared to defend it. When this statement was broadcast, it was a warning against any external sabotage attempts, especially by the Mulla. The above-mentioned Congress investigations record that the US President and his Adviser on National Security had postponed approval to send arms to the Mulla and his faction, despite the recommendation of the CIA representative and the Shah's endorsement. But on the day following the nationalization, a decision explained that the aim was simply to weaken the Iraqi rule, and not to achieve a victory for the Kurds, because a victory for the Iraqi Kurds – as was explained by the US authorities – would affect the political stability in Iran.

To return to the declaration of 1 March 1973. That declaration meant that the Iraqi Revolution has extracted a great victory with the minimum loss. When the declaration was broadcast, the Government was in a position to immediately fulfil the promise made when the masses were asked for a sacrifice. The Government had promised to pay back what was previously deducted from their incomes after passing the critical period. On March 1, it was announced that "the RCC decided to stop the deductions, and to start back-payment of deducted amounts at the same rate and at the same periods of deduction with effect from 1 April".

What was going on? Were the companies a "paper tiger"? In fact they were not. They were a real tiger. But even a real tiger can be killed. According to the Iraqi Minister of Information:

"It is related in the myths that there was a dragon which brought fear to the villagers, who gave him a maid every morning, to appease him and avert his threat of destruction. He remained for a while, until a valiant knight appeared in the village who did not fear death. So, he came to face the dragon and aimed a sharp arrow at his heart. The myth collapsed, and the villagers lived in peace.

"Then came nationalization to Iraq on 1 June, which meant that one of the villagers – one of the oil-producing nations – that had lived for years in the midst of fear and embezzlement, presenting sacrifices to the monopolies' dragon, had risen to his feet with courage and decision and aimed a sharp arrow at the heart of the dragon. Contrary to the myth, in reality the dragon does not die with one blow, but with such a blow his death begins".[4]

This is the story then – and the dragon is not dead yet. In April 1974, mutiny began in the Kurdish area, supported by the US and from across the Iranian borders. Thus exploded the mine which had been kept in check for a long time. But the revolutionary finger was still on the trigger as Saddam Hussein reports:

"The method of imperialism, according to our limited experience, resorts to damming the river down from the spring, if it cannot control the river from the start, imagining that after the spring the flow becomes weaker or needs less effort in attempting to go against the current. America may give the arms, but people who carry American arms are few. Decisive action to remove a pocket of reaction is not decided by arms, but by politics. As much as politics echo the interests of the masses, Kurds and Arabs alike, it can push this pocket to the last position of retreat."

But, as conspiracies were springing up all around, the situation had another soft spot, open to conflict. The nationalized oil, which poured into the Mediterranean, came out of the domain of the Revolution, protected by the patriotic front, across the territory of two states. This fact would undoubtedly remain a threat, usable at any moment.

It was not necessary for either of the two countries to take an open decision to stop the passage of oil. But it was possible that the pipelihes could be "sabotaged". There was no doubt that safe-guarding income resources and the protection of economic independence against conspiracies demanded preparation to face this possible danger.

Saddam Hussein did not leave the question open, but said very frankly:

> "We do not think that imperialistic plans will use the Kurdish issue only... They will use everything, and we also have to prepare in every direction to surprise those plans with unexpected counter-plans. Wherever they act, we have to react with three or four counter-attacks to cause them damage. We do not defend Iraq in Iraq only, and they can never lead us to this situation.... We do not think that we are reminding imperialism of something they do not already know when we say that imperialism plans to use the pipelines, which lie outside our sovereignty. This is something which we expect every day and week and month, taking into consideration all possibilities, facing the incident quietly. We shall not be surprised if we are told one day that they are tampering with the pipeline somewhere, somehow. We shall tamper with twenty of their pipelines. Yet, this alone is not sufficient. We have to decide in advance how to hold out for a long period of time in order to tamper with twenty of their pipelines."[5]

I think that the companies' experience of conflict with the Iraqi system over many years convinced all quarters of the imperialistic conspiracy that the warning made by Saddam Hussein was not mere bluff. Therefore they refrained from taking that serious step. Whether that restraint was caused by estimates or hard information, it was no longer a secret that the threat was backed up by more than words. It was a declaration of a composite plan ready for immediate implementation.

But there is that last sentence in Saddam Hussein's warning which calls for another look. The man had shown another aspect of his plans. He said we shall not merely take revenge by damaging twenty of their pipelines, but we have to prepare the means for a long period of fortitude. This meant that Iraq had to preserve enough assets of foreign currency. In 1974 Iraqi use of oil resources left a surplus of 900 million dollars. Yet, at the same time, Iraq was trying to get foreign loans. When Saddam Hussein was asked about this question, he said clearly: "We make plans for everything. We took the loans out of need not out of whim... but based on present and future calculations. Our revenues pass through oil pipelines, some of which lie outside our sovereignty".

Despite this, all these preparations were temporary security measures. Long-term strategic insurance was underway during that period. In January 1973, after signing the agreement with the Syrian Government, a decision was made by the political Command to start building a pipeline inside Iraqi territory ensuring the export of northern oil from Basra in case of need. The technicians objected to the hasty implementation of this "strategic line" because of the high cost of installation of such a pipeline, at a time when export rates of Kirkuk oil were still far from their normal level. But the political Command refused those reservations, backed up by a broader picture of circumstances.

On October 19, 1973 a contract was signed to implement the project of a pipeline from Rumaila, Basra to Hadeetha. The signing took place at the INOC headquarters in Baghdad, between INOC and the Italian group (Snam-Progetti, Seibem and Montube). The costs were estimated at 44.5 million dinars, with the project to be finished within 22 and a half months from the date of signature. The contract covered all works of construction and the supply of all needs for the project, except the main line to carry crude oil and gas. These were contracted with the Japanese and French group of companies, at a total cost of 27 million dinars.

What is noticeable here is the variation of international relations. Iraq also exercised the policy of an "open economy" but on different lines. The important thing was that the decision was made. The RCC entrusted the chairmanship of the Follow-up Committee on Oil Affairs and Implementation of Agreements with the responsibility of implementing this project. Another project was entrusted to the same Committee: the Deep Water Terminal (Al-Bakr Terminal) in order to sidestep traditional methods which could have held things up.

It was possible in fact to complete the strategic line on the date given. Again it was proved that oil battles were more than technical exercises, which need to take all factors into consideration and lack the courage in accepting calculated risks when taking decsions. Taking oil to the sea became safe from attempts at sabotage or embezzlement. When the Iraqi Government asked the representatives of the Syrian authorities to negotiate for the amendment of the January 1973 agreement, the Syrian authorities asked for a new raise in transit dues from 45 cents to 180 cents. The Iraqi side to the negotiations in March 1976 was able to give a flat refusal! Iraq was also able to stop pumping oil into lines going through Syria and divert all northern oil – through the strategic line – to be exported from the South. Concentrated efforts secured this diversion so that Iraq did not lose a single drop of oil, or a single fils of government income, or any of the basic customers.

In addition to all this, it luckily happened that the early implementation of the strategic line saved scores of millions of dollars in construction costs, because of the high rate of increase of the price of imported materials in the last few years.

## NOTES

1. Private interview with the Iraqi minister of planning (1976).
2. Text of March Agreement, Al-Mousawi, *op. cit.*
3. *Newsweek,* July 3, 1972.
4. T. Aziz, *op. cit.* (chapter on nationalization).
5. Saddam Hussein, *On Current Affairs* (Baghdad, 1974) (Arabic).

# CHAPTER 15

## SUCCESS IN OIL MARKETING

Iraqi sovereignty in the field of oil was settled after the March agreement, and the BPC operations were controlled within a specified framework. Yet, it was expected that all of this was but a transitory stage. And so it was.

On 6 October 1973, the Arab heroic war against the Israeli enemy began, and the ABSP Command took the decision to put all Iraq's strength into the service of the battle. A number of measures were taken, among which was a decision to nationalize the American and Dutch shares in BPC. On 7 October the American share was actually nationalized by Law No. 70 of 1973. These shares (23.75%) were owned by Standard Oil of New Jersey-Exxon and Mobil Oil Corporation. The Minister of Oil and Minerals summoned the Representative General of BPC and informed him officially of the decision. INOC then began marketing this nationalized share.

With this decision on the second day of the fighting, challenging nationalization and according to national political justifications, announced this time, Iraq became a partner in the BPC capital by unilaterl legislation. The companies did not dare to object. With this decision Iraq became a partner in the top level of administration in the London office, and the national administration had to market larger amounts of oil.

On 21 October, another law was issued (No. 90 of 1973) to nationalize the public share of the Dutch Oil Company in BPC (representing 60% of Shell oil). This law brought the INOC share up to 38% of total BPC shares in two weeks. The attack was completed on December 20, with Law No. 101, by nationalizing the public share belonging to the Exploration Corporation owned by Gulbenkian (5%). This company has its headquarters and management in Portugal, which had an openly antagonistic attitude during the October War.

Two years later, on 8 December 1975, a joint meeting was held by the National and Regional Commands of the ABSP, presided over by President Al-Bakr, where it was decided to nationalize the remaining shares of BPC. Then a joint meeting was held by the Council of Ministers and the Supreme Committee of the Progressive Patriotic Front, chaired by Saddam Hussein, where the decision was presented and approved. At 6 p.m. President Al-Bakr made a statement to the people concerning that decision, and the RCC issued Law No. 200 of 1975, to drive the last nail into the monopolies' coffin.

From that day nationalization was comprehensive and complete. Iraqi nationalization does not merely mean the possession or complete state control over production operations in the fields, it extends to all areas. The Iraqi state alone plans the development of oil production and the processing of a growing proportion of that oil. Iraq alone decides the method of using the income of oil export in comprehensive development and a planned national economy. The Iraqi state confirms economic independence, and renounces foreign attempts to cause pressures limiting the freedom to issue political and economic decisions, through state policy, to market Iraqi oil.

We shall now postpone dealing with some of the previous points, and consider the last point, which is actually the starting point of a new issue, that of independent marketing. The Iraqi Command decided from the first day of nationalization to take direct charge of marketing by commercial contracts, without mediation of any sort by the monopolies which exercise an octopus-like control in oil markets. After the 1975 nationalization, the state responsibility for marketing came to cover all exports of Iraqi oil (in 1974 reaching 92 million tons). Control had never been exercised by any oil-exporting country on such a large scale. No national company, even in the Soviet Union, had embarked into world markets with such a large amount of oil. Boris Ratchkoff used to say that Seyus Naft Export (the Soviet Oil Export Establishment) was the only independent exporter in the international oil markets controlled by American and British monopolies.[1] But when INOC undertook independent marketing, a larger giant stood beside Seyuz Naft Export since the total Soviet oil exports to Socialist and non-Socialist countries did not exceed 70 million tons in 1970.[2]

Naturally, what facilitated marketing was the mutual understanding among the OPEC countries. After the historic successes of these countries in setting a just price structure, it became necessary to carry this mutual understanding another step forward to cover programming of production and to protect price structures. The Iraqi state took the initiative to suggest that, and called in 1974 for a meeting of the producing countries which had national companies. After preliminary bilateral contacts, the meeting was actually held in London early in 1975, hosted by the Iraqi Government, and under the auspices of

OPEC. The Iraqi point of view called for an agreement about a programme of oil production, keeping a general balance between supply and demand in the world market, provided the demand be distributed to various supplying countries according to objective criteria.

There is no doubt that the Iraqi demand came in time. The economic recession in the consuming industrial countries, and the measures taken by the International Energy Commission led by the United States caused a decrease in demand. "When a decrease in demand of 4.6 million barrels a day occurs for instance, then," as the secretary of the Follow-up Committee said, "in order to avoid probable harmful competition among various producing countries, we have to co-ordinate within the framework of OPEC to distribute that decrease on reasonable bases."[3]

The temporary surplus, fabricated to a large extent, was a weapon in the hands of the Energy Commission and the oil monopolies intended to cause a split among OPEC countries and stir competition by raising the export rates of this country or that, without a reasonable basis, but according to their political climate. Adopting the Iraqi proposal to set one's house in order away from antagonistic manoeuvres was a way to enhance the role of OPEC by protecting the price structures. In the London meeting, Iraq submitted a number of criteria to limit each country's share of oil production. These were: population of the country; real need for finance development; size of foreign currency surplus; the country's oil production capacity (size of reserve); and removing historic injustice done to some countries as a result of the monopolies' conspiracies.

There is no doubt that these principles were objective and useful to all parties. For what was the point of any country raising the rates of oil production if that country had a limited capacity for revenue intake? Such a country would be doing nothing more than exchanging a real wealth of growing value for money in foreign banks decreasing in value. The question becomes more eccentric when there is another country of a higher intake capacity, in which raising the oil production rates could reflect directly and fully on serious economic development in that country.

Commitment to any criteria agreed upon was useful to all parties, and the exporting countries as a whole, and it was ultimately helpful to the Iraqi Command in marketing oil in quantities and at prices which agreed with national interests.

In fact, this was not a new idea, since it had been discussed by OPEC more than once. In the ninth conference in Tripoli, Libya, in 1965, an attempt was made to lay an experimental plan for the following year. The conference recommended that the rate of increase in Saudi Arabia and Iraq should be 9%, Iran 14%, Kuwait 5% and Libya 33%. But Saudi Arabia did not attend, and objected to the resolution, so the actual rate of increase in Saudi oil jumped to 18% in 1966. Shaikh Yamani stated at the time that the question was not that of programming, but of economic factors and resources.[4]

While the idea was not new in principle, it was thought that the development of OPEC could provide a better chance of understanding but the Iraqi initiative failed in the first meeting and the issue was lost. Undoubtedly, this failure alone exposed OPEC to serious dangers in the following two years; and it was another difficulty facing the marketing of Iraqi oil.

Now, what had the Iraqi national administration achieved in the field of marketing after it became the full responsibility of the state, despite the fall in demand in world markets, and despite the OPEC members' refusal to co-ordinate?

Concerning the size of export, we may remember the spiteful remarks of Shaikh Ahmad Zaki Yamani, the Saudi Oil Minister, in referring to the fall in export rates a few months after nationalization. In fact the total export of Iraqi oil was about 78 million long tons in 1971, but it dropped to 67.3 million long tons in 1972. But, in 1973 it reached 93.2 million long tons. For the Kirkuk oil in particular (the Iraqi Company for Oil Operations) the export was 33.1 million long tons in 1972, but it went up to 52.8 million long tons in 1973. In 1974 another fall in the total export occurred once more (89.6 million long tons) despite a rise in exports by BPC and INOC, owing to a sizable fall in the production of the Iraqi Company for Oil Operations, exporting only 40.2 million long tons[5], a fall of 23%. An Iraqi official says: "This fall in production was international, and by choice, in order to protect oil rates against supply and prices, by unilateral action. When we were left alone we decided to change our policy in a way that related oil production to development needs, while continuing a close observation of supply and prices". The fall of production in 1974 did not affect the income because of the increase in oil prices according to OPEC decisions in late 1973. Some estimates say that Iraqi income from oil exports rose from 1900 million dollars in 1973 to 6000 million dollars in 1974[6]. In 1975 the picture was different: a fall in total demand in the world market, leading to a fall in rates of oil production and export in all producing countries except one, Iraq.

In the first half of 1975, the world production (outside the Socialist bloc) dropped by 11.4% (in comparison with the same period in 1974). The tendency continued to the end of the year, reflected in a fall of production in the State of Arab Emirates by 5.2% during January-September, 1975 (compared to the same period in 1974). In Algeria the fall was 21.9%, in Saudi Arabia 15%, in Qatar 20% in Kuwait 17.4 and in Saudi Arabia 15%, in Qatar 20% in Kuwait 17.4% and of 16.7%[7].

How did this happen?

The official responsible for Iraqi oil marketing gave this explanation: the Iraqi marketing policy starts with a clear conception of the world oil situation and the movements of antagonistic forces. We have enough information on market developments, according to which the Iraqi state exercises its oil marketing. We have the independence to lay a sound strategy in oil production and marketing, which is in tune with our national interests and political principles. The official adds that some quarters claim that the rise in the Iraqi product is achieved at the expense of price structure, but they do not present a single shred of evidence to support this claim. They rather know that our marketing principles are stable and internationally known. But they encourage this claim to justify their own failure in marketing, which is partly due to their relations with the concessionary companies.

The other countries, with no exception, still retained some degree of contact with concessionary companies in marketing their oil. These companies received the oil at reduced prices (22 cents per barrel less than market prices, a reduction called "company incentive" to buy from the state. Incentives were used by all: it was found in the Iranian Consortium, in Saudi Arabia and Kuwait, in Algeria and in Libya, though in different proportions. There is no doubt that the absence of full control over marketing operations in these countries ensured that the future and consequences of marketing would be controlled by one large monopoly, planning on an international level, gaining power from its giant size and long experience, and desire to secure the highest profit margin, and concern to stay the longest period possible in the producing countries. All of this could not be achieved except by allowing these countries to feel the weakness of their position and their inability to market large quantities in a market where there was a certain recession. All the measures taken by the companies, in collaboration with the International Energy Commission, and the ferocious psychological war exercised by every possible means, ensure the objectives of those companies. The intricacies of marketing were outside the reach of the producing countries, which then become at the mercy of the market "giant" which controlled resources and destinies. And the weak countries were consequently at the mercy of the companies to market any amount, for which more "incentives" had to be offered. These countries derived their information and experience of the market situation from biased sources, that is, the concessionary company buying the state oil. And, whoever possessed the information possessed control, and could inspire with terror the local technicians by leading them to believe that their oil was unsaleable. At such terror, the producing country remained under the control of the giant concessionary company, despite any change in form of ownership.

All this was put by the Iraqi official to the Kuwaiti Oil Minister: He said:

> "Why do you limit yourselves to one company? Spread your oil over the world and you have a complete feel of the market. When I am limited to one company, the information and experience reach me from one source. But when I spread, I can judge any idea through various perspectives and data, hence I can realize my exact position. Naturally, the Kuwaiti authorities did not listen to the advice, and decided instead to decrease the price of their oil, delivered to the companies, by 10 cents, by taking a unilateral decision* which affected the entire Gulf oil."

This then was the first point in the answer given by the Follow-up Committee: the others were responsible for threatening the oil price-structure, by submitting to conspiracies and the embezzlement of companies which undertook their oil marketing.

There was another point in the Iraqi reply. When all the oil countries agreed to sell large quantities of their oil to the giant concessionary companies by reducing the price, or by incentives reaching 22 cents, becoming 32 cents per barrel with the Kuwaiti decision, "Why then all this noise about Iraqi reduction of oil prices? Why is it demanded that INOC stick to OPEC prices and thus be deprived of flexible movement available to Gulf Oil exploitation, receiving oil at 32 cents less per

* The reference is to the decision of the Kuwaiti Government to decrease Kuwaiti oil prices in November 1975 contrary to the unanimous decision of the OPEC Ministerial Conference. A statement by the Iraqi Ministry of Oil and Minerals commented that the Kuwaiti measure came at a time when OPEC unity was a target of attack organized and co-ordinated by various quarters in order to damage the price structure and destroy OPEC.

barrel? INOC is no longer a small local company like Kuwait Oil Company or the Saudi Petromin which cannot market more than 10 million barrels a year. INOC sales are now larger in size than sales of Gulf Oil and Shell put together." Yet, did this mean that Iraq used this as an excuse to threaten the price structure, as claimed by its adversaries? The Iraqi official confirms that "though we have the right to be treated like a company, and thus reduce our prices by 22 cents, and enjoy the flexibility available to such companies, to maintain our position in the market, yet we have preserved the price structure.

Now we move to a third point. There is a reasonable question that one can ask at this point: was it in the interest of Iraq to reduce prices? Naturally, preserving the price structure was in the interest of all the OPEC members who had reached the present price levels after a difficult struggle. But the Iraqi Command in particular, with its general revolutionary policy, and in the light of its complementary oil policy, had to be most careful of all to preserve prices, and the very last member to aim a blow at the achieved price structure, even if some of the others might have to do so. Why? Because the Iraqi Command realized that such an independent oil policy would arouse the highest degrees of rancour and enmity from monopolies and imperialistic powers. The fall of the OPEC front would certainly threaten the entire Iraqi future more than any other country. The Iraqi Command had to make every endeavour, in self-defence, to support OPEC unity and develop OPEC attitudes in all fields. In the question of price structure in particular, the Iraqi state had made a distinguished contribution, but that is another story. So, how could Iraq be considered to have an attitude contrary to strategic interests and become the first member to destroy the price structure? Moreover, the marketing policy drawn up by the Iraqi Command depended on commercial contracts, and deals with national companies in consuming countries directly, without an intermediary. This meant that Iraqi oil sales were offered in the "free" market, and that they did not hide behind giant companies dealing with markets under their control, who could, ultimately, secure a minimum price of crude oil owned by the producing country. To support the position of Iraqi oil in the free market, and in facing big consumers, demanded first and foremost a solidarity amongst the producers. This was the basic defence line of the Iraqi oil price. If this should fall, Iraq, facing the market alone, would be dealt the severest of blows, which would not be comparable to any damage which could affect the others.

It is obvious then, Iraq could not possibly take a step to destroy the price structure.

And here comes the fourth and decisive point announced by the Follow-up Committee:

> "We said at the OPEC meeting that we were ready to display all of our contracts before you, provided you too display all your contracts on the table. But the others refused. They do not have any evidence to support their charges. At the same time we have very precise information about large deals harmful to the economy of these countries. We have evidence about very cheap contracts by the same countries which accuse Iraq of selling oil at a cheap price."

This was a refutation of charges laid against Iraqi marketing policy, but the motives for propagating these rumours remained a source of real fear, as they had political implications. It seemed that the strategem of the International Energy Commission and monopolies had succeeded. It also seemed that there was a tendency to reduce oil prices in response to U.S. policy, a tendency which was looking for a justification.* Was there a stronger justification that could be made public than saying that revolutionary and stubborn Iraq was exercising and starting the game of price decreases?

Before the nationalization of Kirkuk oil, all Iraqi oil, except the limited quantities of direct development oil from Rumaila, was delivered to the monopolies for marketing. The role of the Iraqi Government, as was the case with all other producing countries, did not exceed that of a tax-collector (government revenue or royalties). The Government used to get so many cents per barrel, and there the matter ended. The concessionary companies in Iraq used to market 90% of product in the U.S. or some European countries, where there were large branches of the companies sharing in IPC or BPC.

After the Kirkuk nationalization, and then after the comprehensive nationalization, the connection of concessionary companies with Iraqi oil was completely severed. Marketing was no longer subject to strategies by these companies, but to the strategy of the Iraqi state. From the beginning of marketing in 1972, the basic principles of this strategy were laid, then they were crystallized and developed with

* Material for this chapter and the entire book was collected and written in the first half of 1976, before the split inside OPEC about prices which happened late in that year.

experience. That strategy may be outlined as follows:

Firstly, oil was marketed without an intermediary, i.e., by direct contact between owner and consumer.

Secondly, the buyer had to be an oil processor and not a trader, i.e., the buyer should own a refinery to refine crude oil before going to the final consumer. This prevented speculation which offered the oil of a certain country in the same market for more than one price. Speculations do not follow set economic principles, and they do not follow mere supply and demand, but a number of factors which enter into the form of trade relations among countries. It is known for instance that exchange between any two countries may follow the barter system or deferred payment, and this somehow affects the price of the commodity exchanged. If it happened that a certain country should re-export Iraqi oil in this manner to a third party, that country would be speculating with imported Iraqi oil through INOC, and the same oil would have more than one price in the same market. This is like finding a car in the market at a price fixed by the State Automobile Company, then finding the same car with the same specifications displayed by another merchant at another price. There is no doubt that this situation would affect the decision of the buyer and make him hesitate, since the consumer prefers to buy a commodity of a fixed price.

Thirdly, the export policy was to approach the oil processor and avoid the oil traders, then this basically led to the situation where the national companies undertook these jobs of importing oil to be processed for local consumption in those countries. Between INOC and such companies there were joint interests: neither belonged to the international cartel, one being a national producing company, the other a national consuming company. Both exercised their duty while facing pressures from the giant monopolies. This made dialogue between the two parties objective and generally reasonable, free from attempts to cheat and manipulate, as was the case between a weak party like the producing country and a giant party like the monopoly. Moreover, the national company remained as long as its country remained. In that sense, the relations between the national producing company and the national consuming company were of a stable nature, both seeing to the continuation and stability of the other party, since the national producer could not do without the national consumer, nor could the latter do without the former.

The fourth point in the strategy was to spread in the largest area possible. This variation in relations saved the Iraqi state from attempts by outside parties to apply economic pressures. If all the oil were to pour into one country, then any delay by this country in buying Iraqi oil would become a painful means of pressure, ultimately affecting the independence of the political decision in some degree or another.

The Iraqi marketing policy has achieved this aim. Official statistics for 1974 show that the west European share was 63.2% of total oil exports, North America 1%, Latin America 8.8%, Africa 5.8%, Asia 10.8%, Socialist countries 8.7% and Australia 1.7%[8]. These proportions give a general indication of the size of the spreading market, and the detailed figures of each country are supposed to define this result more precisely. But the amounts of export to each country have now become unpublishable national security material. Yet, it is enough to say that countries to which oil is exported now are: (western Europe) Greece, Italy, France, Spain, Austria and Belgium; (Socialist Europe) the Soviet Union, Bulgaria, Hungary, the German Democratic Republic and Poland; (Arab countries) Egypt, Sudan, Somalia, Maghreb, Syria and Lebanon; (Africa) Uganda; (Asia) Turkey, India, Sri Lanka, China and Japan; (Latin America) Brazil; and Australia.

It should be noted that this distribution map extends to various continents and political blocks. Yet the marketing administration tries to open still more markets. It should also be noted that this large number of contractors gives each country an average of 5 million tons of crude oil. This is merely a confirmation of the plan not to fall into one hand. But in fact there are relatively large buyers on the list including France, Italy, Japan, the Soviet Union, Brazil, Turkey, India and Syria.

There is also economic value in this extension, alongside the political value. From such a large area the marketing management collects various data directly from oil buyers, who represent their governments, and some large national companies. This profusion of information is one of the basic secrets behind the success of Iraqi oil marketing.

The next point in the strategy was the pricing of the Iraqi oil according to rules followed in pricing oil within OPEC. This applied to Kirkuk oil sales from the Mediterranean, as much as it did to Basra oil sales from the Arab Gulf. The Iraqi Command was quite aware of the fact that Iraq would be the first loser when the price structure collapsed. The Command realized that all sales were done according to commercial contracts, and were not secure like the sales agreements with the concessionary companies.

The strategy also allowed for controlling production rates in a way that avoided excess in supply, and consequently a fall in price. Supply was limited by a precise balancing, as far as possible, with demand on Iraqi oil in the market.

On the other hand, balance must also be precisely maintained between production size and the need for necessary income to provide development plans. This included agreements on financial and technical co-operation held in the period 1972-5 with a large number of countries where oil was a basic item, in return for installation of the best technical equipment in Iraq, at the lowest possible costs, as well as the necessary technical education. It is not true that these dual commercial agreements insisted that Iraq should buy certain commodities in return for oil exports. Agreements are usually arrangements of payment in return for deals of independent imports, and in accordance with prevailing world prices.

The strategy also catered for the fact that among the principles of marketing there should be the preference for countries sympathetic in their political attitudes towards the national issues. In the 1973-4 period, when consumers were terrified, the Iraqi Government opposed the random decrease decisions. Iraq completely prohibited oil exports to the U.S., Portugal and Holland (and nationalized their shares too). But Iraq was careful at the same time to assure the countries which had a friendly or neutral attitude. This explained the principle which controlled the Iraqi marketing policy to various countries. Such a policy could not put all countries without distinction into one camp antagonistic to the Arabs, and direct all punishment on to them.

All the previous strategic points could turn into mere "big talk" without the following principle. This one lends great weight to the previous principles and, at the same time, ensures the transformation of theoretical principles into successful practice. And this is, the power entrusted to the chairman and Secretariat of the Follow-up Committee, and the methods followed by both. Comrade Saddam Hussein and the Secretariat had full authority and flexibility to assess the significance of every contract with a consideration of aims to be achieved. Running the operation with routine office mentality would have been devastating when dealing with hundreds of millions of dollars, and in a market which set a high value on time, where no mercy was afforded. Undertaking such operations needed a mentality that knows no vacillation or fear, with the ability to take the suitable decision with confidence. This practice sounds easy to talk about, but it is difficult to apply, since it calls not only for self-confidence, but for the confidence of others in the man with the power.

Concerning the methods of administration, the seious decisions went through a very short process to reach the man in charge. Longer processes lead to negative results. The larger the administrative system, the smaller the ability to take fast decsions. Also, a limited administrative system is easy to supervise, where in this type of business any mistake, however small, should be traced and confined and responsibility verified with speed and precision. But on the other hand, a limited number cannot deal with the large amounts to be achieved, which call upon the system of administration to achieve the highest productivity possible. The Secretariat tried to achieve a version which ensured the smallest number and the highest efficiency through personal experience and information garnered about the development of administration systems in advanced countries.

The choice of personnel came next. Efficiency alone was not enough in this field, since everything connected with oil was also connnected with national security, and complete secrecy was imperative. If the price was disclosed for instance, the amounts or terms of a contract with country X became known and flexibility of action with other parties became very limited. Methods of action also had a certain nature which should not be known by others. This was a part of the know-how resulting from long experience, and from the personal qualities and way of thinking of the decision-makers. It was also the result of calculations undertaken by different methods than those used by other countries. And all this was a secret.

The last strategic principle followed is related to the previous one, or it is rather a result of it, but it has become an asset in the international market. This is the good reputation that Iraq has acquired. The risks in oil marketing are fearful. A loss of one cent per barrel means a loss of $75,000 in a one million ton deal of oil. By the same standard, a drop of 10 cents leads to loss of $750,000. The Iraqi exports are about 100 million tons. Consequently, sharp supervision and strictness in choosing personnel limited, if not entirely prevented the temptations of bribery and commissions, etc. These matters would not only harm the Iraqi treasury, they were also harmful – which is more serious – in the leakage of information. The Iraqi marketing system over the past few years has acquired an international reputation as a clean system accepting no bribes or commissions, and enjoying stability and clarity in contracting – no intermediaries, no third person between seller and buyer. In some countries you may find an office director who contacts a purchaser to say that he can arrange things, then another official who contacts the same party to offer a lower price if a deal is to come through him, etc... Western newspapers publish many such scandals of the oil world. Stories have been published about oil buyers who sign contracts with some countries, and on the plane they receive a telex to inform them that the deal has been cancelled. None of these stories was found while dealing in Iraqi oil. The marketing leadership was invulnerable, having Saddam Hussein in command. This leadership was careful to strengthen this reputation and to establish a tradition of confidence. In January and February 1974, the oil prices went up to $22 per barrel, and these traditions were put to a severe test. It was possible then for the Iraqi state to alter all previous contracts to bring them up to the new market price. Long arguments took place on the subject, as the Iraqi treasury would have gained a lot from altering those contracts. When the Follow-up Committee submitted the outline of argument to Saddam Hussein, the answer was decisive: "We are concerned with future relations and not with immediate gains, therefore contracts remain as they are, and prices are not changed". All buyers in countries and national companies and their consultants remember this attitude and consider it a safety-valve.

When purchases of Iraqi oil by country X reach $500 million, we can imagine the size of loss in that country if the price was altered. Since oil is a commodity of special nature, influencing the entire economic movement, then this would double the concern of those countries to ensure stability in dealing. This ensuring of stability remained the basic excuse raised by the cartel companies to convince the consuming countries of the importance of those companies. In fact the sense of confidence and stability form a vital factor in all oil relations. Owing to their stability in oil supplies and prices, it became known that INOC contracts had a high rate of implementation all the time. Preserving this high rate was due to the mutual care of both the Iraqi side and the purchaser.

So Iraqi oil now enjoys a reputable character. And this "commercial name" opens the markets and spares us the game of price reduction. A great number of purchasers prefer dealing with exporters who ensure regular supplies and stable terms to running after transitory speculations or cheaper but irregular deals. Prices are not the only factor in deciding the size of sales. A country like France had previously explained that her policy in the field of oil is summarized in "the ensuring of safety of supplies and variety of resources, even when this is not achieved at the cheapest prices". The French attitude is not an exception in this field.

When the Iraqi official finished his wonderful display of the ten principles of the oil marketing strategy, I asked: "Why does Iraq not offer this experience to others? Do you ever think of helping others to market their oil?"

The Iraqi official answered:

> "This happened once, when we said to Kuwait before they reduced their oil prices: 'If it were said to you that there is a surplus on the market and that there is no desire for Kuwaiti oil, can we then enter into the marketing operation'. Naturally we received no response, and we are not thinking now of making the same suggestion to other countries. The offer may be wrongly received, and there are sensitivities that must be observed."

**NOTES**

1. B. Ratchkov *op. cit.*, p.264.
2. Q.A. Al-Abbas *Oil and Development* I,4.
3. *Oil and the World* XXVIII, October, 1975.
4. Sampson, *op. cit.*, p.165.
5. Official statements of the Ministry of Oil.
6. M. Field, *op. cit.*, pp.52.
7. For figures of oil exports of Arab countries until September, 1975, see *Oil and Arab Co-operation,* II,1, Winter 1976, (OAPEC Publication).
8. OAPEC, *Second Annual Report,* 1973-4 (1975).

# CHAPTER 16

# OPEC'S STORY

In the previous chapter, I have written extensively about the price structure achieved by OPEC. This calls for a look back at a near-by, though not well-remembered, date, in order to answer the question: How were the historic decisions about price made in October and December 1973? Really, is it not high time the story were published about these historic decisions, and about how they were come to inside OPEC?

The story begins with the fact that 6 October witnessed the beginning of military movement, and also the movement of a team of oil companies' delegates to meet in Vienna and start talks with OPEC about new prices for oil. When the talks failed on 11 October, the news about military action was occupying the front pages, and the news of failure passed without much concern on the part of the Arab masses. In fact, the news about those talks was treated lightly everywhere. The western newspapers moved the news about the failure of the OPEC talks to the business pages. Even in the *Financial Times,* the news was in a corner of page thirteen.

But the failure of the Vienna talks was not all that unimportant, and did not deserve to be thus ignored. The delegations of producing countries returned home, and an emergency meeting was announced, to be held in Kuwait on 16 October.

On the evening of that day, Dr Abdul-Rahman Khan, OPEC secretary at the time, announced to the world that the countries in the Kuwait conference had unilaterally decided to raise prices of Arab Gulf oil by 70%, and North African oil by more than 10%.

The four corners of the globe shook at hearing the news.

No doubt the atmosphere and the self-confidence propagated by the October War created an incentive to challenge the imperialistic forces, and encouraged some oil-producing countries to take the risk. But, beside this, there had been an acceleration in confrontation adopted by OPEC, years before that historic day. It was natural, then, that the situation would come to a head one day, under suitable circumstances. And the October War was the historic moment for this explosion.

Till now, I have added nothing new to the known facts. But it should be mentioned that on 12 October, a statement was made by the Ministry of Oil and Minerals in Iraq saying that "this decision was the outcome of long and tiresome efforts made by OPEC towards this objective [i.e. limiting posted prices] by state control of oil wealth and exercising sovereignty effectively, especially determination of the oil price, and then the revenue of sales and exports of oil". The statement added, "Iraq has played a very important part in the achievement of this development". This sentence passed unnoticed, and reports of the October decision did not pay attention to it.

On the contrary, attention was directed elsewhere. The story is told about a team representing 20 oil companies carrying previous instructions from London (the capital of oil companies) saying that a price increase should not go beyond 25%. The Iranians wanted the price per barrel to go from $3 to $5. But Shaikh Zaki Yamani, the Saudi Oil Minister, began by asking for the price to be doubled.

On 10 October, while the war with Israel was at its peak, the companies' delegates cabled to their superiors, who were waiting at the headquarters, to say that both parties were still very far from any hope of agreement, but that there was a possibility of reaching an agreement about a price somewhat less than $5 per barrel. Esso and Shell answered by saying that the companies must inevitably stop there, and Shell declared that the financial consequences of such an agreement called for consultations with the governments.

After midnight, the story goes on, George Percy telephoned to Shaikh Yamani to say that the companies needed a two-week period for consultations with governments of consuming countries. Another story says that Percy (Esso) and Bernard (Shell) went to see Yamani at the Intercontinental, to inform him of the time needed, and that Yamani meditated for a few minutes, then called up his assistant Dr Ibrahim Ubaid and started discussing with him preparations for returning to Saudi Arabia. A third story says that, after meditating for a few minutes, Shaikh Yamani ordered a Coca-Cola for Percy and went on quietly squeezing a lemon into the glass, to gain time, hoping that Percy might change his mind and resume negotiations. This third story adds that Yamani telephoned to Baghdad and spoke Arabic, then turned to the representatives of the companies and said: "They are extremely angry with you". Then he began turning over the pages of flight schedules. When it became obvious that the companies had nothing new to offer, the Shaikh left, and called for the Kuwait meetings, and the decisions were made.[1]

Does this story, with all it implies, suggest the real history of the October decision? We shall see.

We return to the Tehran Agreement of February 1971. This was considered an important event at the time, and the OPEC negotiation team was composed of Saudi Arabia, Iran and Iraq. In that agreement, new prices were set to reflect the growing rates of inflation. It was acceptable at that time that expected rates should be 2.5%. The agreement also fixed a new method of calculating taxes and a system of pricing. The Tehran Agreement was understood to cover long distance oil, therefore it was followed by complementary agreements to cover short distance oil (the Tripoli Agreement to fix the price of Libyan oil, the East Mediterranean Agreement to cover Kirkuk oil, the Saudi Agreement to cover oil exported from Sidon, the Lagos Agreement to cover Nigerian oil).

But, after the Tehran Agreement, there was a devaluation of the dollar. (It should be noted that posted oil prices are given in dollars, and that a great deal of producing countries revenue is also calculated in dollars.) Therefore OPEC entered into new negotiations with the companies in Geneva, and OPEC was represented by the same negotiation team as that of the Tehran Agreement: Saudi Arabia, Iran, and Iraq. The negotiations yielded the so-called First Geneva Agreement on 20 January 1972 which was considered a part of the Tehran agreement to protect the oil prices agreed upon.

But, when a second devaluation of the dollar took place, it became clear that this Geneva Agreement was not enough to protect the revenue of the governments concerned. So, Iraq called for an emergency conference by the OPEC countries to revise the Geneva Agreement. The Iraqi point of view was that the Geneva Agreement had given a fair increase in prices to match the first dollar devaluation. But it became apparent that the mechanism agreed upon to face future probabilities was insufficient to protect the revenue of producing countries. According to this mechanism, the OPEC countries did not get enough compensation for losses emanating from the second dollar devaluation.

But the Iraqi attitude was met by strong opposition inside OPEC. Most of the countries, especially Saudi Arabia, did not agree to the principle of revising an agreement which was still fresh. The excuse was that the mechanism in question had been previously discussed and accpeted by the OPEC countries, hence the agreement had to be respected and implemented until the date of its expiry. The Iraqi delegation, backed by the countries of non-conventional thinking, said that the agreement is only valid so long as it is fair. When it is found unfair, it has to be revised.

The discussion ended by giving an opportunity to the members who called for revision to present their case. The OPEC negotiation team was formed of Libya, Kuwait, and Iraq, where the latter played a key part by presenting a basic study analysing the shortcomings of the Geneva Agreement and concluding that the agreement could no longer cover the losses. After several rounds with the oil companies, the negotiations succeeded, and the agreement was amended to render compensations more just. Then all the Gulf countries signed the agreement on June 1, 1973 in Geneva, and the new agreement came to be called the Second Geneva Agreement.

During the negotiations, there were basic developments in the oil industry, and the Iraqi Command had been watching these developments very closely, and preparing for a new move beyond the mere amendment of the First Geneva Agreement. And their historic move began only five days after the signing of the Second Geneva Agreement. This move culminated in the decision issued on the evening of October 16.

On 6 June 1973, Baghdad sent this telegram:

Dr Khan, Secretary-General OPEC
VIENNA

We suggest the inclusion of the following paragraph in the agenda of the tenth regular meeting of the conference:

"Review and discussion of Tehran Agreement terms in the light of present and future tendencies expected in the oil market, and the growing increase in prices of processing industries products."

We are sending a report on the same to the Secretariat to be at the disposal of the conference. Kindly pass the suggestion to the member states for perusal and approval. With my best wishes.

Dr Sa'doon Hamadi
Minister of Oil and Minerals.

A few days after the telegram, Baghdad sent the report mentioned by Dr Hammadi in his telegram. The report comprised 23 pages. The title was:

An Inspection of the Tehran Agreement in the light of recent economic developments and the oil market situation. To be presented by the Iraqi Delegation to the Conference No. 34 June 1973.

On p.6 of the preamble to the Report, we read:

"Nowadays the market situations and long range inflation rates have changed, so that the Tehran acceleration has lagged hopelessly behind. If we want to preserve the real value of the posted prices according to the Tehran Agreement, then a new balancing price must be set according to the present market situation, and the present rates of inflation. Taking this into consideration, we have studied the conditions prevailing in the products and raw material markets, alongside of inflation rates."

As a result of this study, the report concluded by recommending a raise in Gulf oil prices of 70%.

But the Iraqi initiative was met with strong opposition. This time the opposition was stronger than the previous one which called for a revision of the First Geneva Agreement. The thirty-fourth meeting of the conference was held in Vienna between June 27 and 28, 1973. The Iraqi initiative was on the agenda as a special item. This was studied in the first session, and in a closed meeting without recording of the minutes, and with no attendance except the heads of delegations. The session extended from 7 to 11.30 p.m., and was resumed on the following morning, again behind closed doors.

The first session began with a sharp dispute. The Iraqi Oil Minister insisted on discussing the subject and taking a resolution, and was backed by a number of ministers present. Shaikh Zaki Yamani headed another group which rejected the question in principle. Dr Sa'doon Hammadi again stressed the Iraqi viewpoint, which stated that the fairness of the agreement was the basis for its respect, and proved, through the Iraqi Report, that the Tehran Agreement was no longer fair, and consequently it was necessary to contact the companies. The conservative group returned to the excuse that signatures of the states had to be respected, and consequently the Tehran Agreement had to continue as valid till the expiry date specified, and that was 31 December 1975. The group said that approval of the Iraqi demand would make future agreements liable to constant change, which would lead to loss of stability in international dealings.

With the continuation of discussion, it was possible for the members present to reach a compromise presented by the Iraqi Oil Minister in the second session. This suggested the formation of a working group to analyse and review the terms of the Tehran, Tripoli, and Nigerian agreements. The group was asked to submit a final report on their findings to be discussed in an emergency meeting, to be held before mid-September 1973. The approval to this suggestion was an important diplomatic success in facing the internal controversy which had begun sharply.

The approval to the suggestion meant that Iraq had succeeded in attracting the attention of its OPEC companions to the fact that there was a serious problem deserving discussion and the right to remain on the agenda. But this success was not only due to diplomatic efficiency, for if that efficiency were not backed up by a successful oil policy inside Iraq, it would not have succeeded.

The Iraqi diplomacy succeeded in convincing all its OPEC companions because it basically represented the nationalization battle with all its challenges and manoeuvres, and because it was backed up by a large amount of direct experience in dealing with the oil market, and because it depends on a national staff of recognizable efficiency, with experience of negotiations with companies. All this had been proved in the negotiations with companies in Iraq, and also in negotiations through OPEC.

Incidentally, it was said during the discussion that the companies could insist on the legality of the Tehran Agreement and refuse any amendment till the expiry date of that agreement. The answer of the hawks – Iraq and supporters – was that oil pricing was a right of the commodity-owner, in the light of the economic situation, and not a question of negotiations. Consequently, it was the right of the OPEC countries to issue a unilateral decision to secure their interests, if the companies should refuse to adapt prices to the variables in the oil market and international economics.

Naturally, there was no unanimous agreement on this measure at the time, and it was not expected. Following the principle of gradual acceleration in such situations, to secure a unified approach, it was necessary to contact the companies first, to prove to all others that they did not respond to the interests of the producing countries – although there was no agreement about contacting the companies which was the suggestion presented by Iraq at the start; and it was necessary to make preliminary studies prior to the contact. Again, following the principle of gradual acceleration, the agreement to implement this preliminary step was a success, in fact an important success, as has been explained.

The action group was formed and was headed by Dr Fadil Al-Chalabi, then the Under-Secretary of the Ministry of Oil and Minerals. In fact he was not new to this sort of work, since he had always headed the technical committees, and the economic committee, since the beginning of the negotiations to amend the First Geneva Agreement.

After this success in the first round inside OPEC, it was necessary for Iraq to continue the activity of convincing the other members. The Iraqi activity took place on two levels:

The first level was the use of direct contacts. Dr Fadil Al-Chalabi was sent to some of the countries by the Iraqi Minister of Oil. It was natural to concentrate on the country of great oil wealth, and also of a cautious attitude. There is no doubt that extensive discussions and clarifications were an important factor in shaking off some of the reservations.

The second level was through the working group which was formed in the OPEC meeting. The Committee of Experts adopted the Iraqi study as a basis, then introduced some developments, to come to a conclusion similar to that of the Iraqi paper, and recommended that OPEC enter into negotiations with the companies for the amendment of the Tehran Agreement.

After this, the emergency conference was held (OPEC 35) in the first half of September as was decided. The Iraqi head of the working group presented his report which filled, with his discussions, more than ten pages of the confidential minutes of the meeting. It was clear from the development of the discussion that the opposition was still maintained though with less enthusiasm. It seems that the opposition at that stage centred on the rates of suggested rise in price and not on the principle itself. It was said that the price suggested by the Committee would shake the world economy, and that the conclusion of the Committee in this respect could not be relied upon as basis for negotiations with the companies, and it was said that it reflected radical tendencies.

The Iraqi Under Secretary reports: "Struggle inside OPEC was great, but the Iraqi objective was clear: patience and pushing from inside OPEC to the utmost limits possible to protect the rights of producing countries, yet with more care to protect the unity of the organization".

In any case, discussions ended with a decision issued by the emergency conference, calling on the companies for negotiations. A ministerial committee representing OPEC was formed for negotiations with the companies. It was headed by the Saudi Oil Minister Shaikh Zaki Yamani and the Iraqi and Iranian Oil Ministers were associates.

Before this negotiation on the ministerial level, there were negoitations in Riad on the technical level, between representatives of Gulf producing countries and representatives of the companies. Dr Al-Chalabi was again a spokesman for OPEC, and these preliminary discussions proved to the companies two things:

Firstly, that the prices of the Tehran agreement had dropped extremely low, in comparison with current prices. This meant that the companies were making extra profits, economically unjustifiable. The general understanging of the Tehran Agreement indicated that extra profits should go to the producing countries, since the companies were merely institutions working for the governments. The growing demand over supply must be reflected in oil price increases, which increases should not go to the companies but to the oil-owning countries.

And secondly, the annual increase, specified in the Tehran Agreement to meet inflation results, was no longer compatible with inflation rates prevalent in western Europe. The OPEC representatives' studies showed various inflation rates which proved that the real annual inflation rate was 6 to 10% and not 2.5% as mentioned in the Tehran Agreement.

The arguments used by Gulf States representatives in the technical discussions crystallized the efforts of the previous months. They concluded by saying that the Tehran Agreement had lost its validity owing to the price itself, and to the inflation rates. The companies' representatives were defeated by the data presented.

When the negotiation went to Vienna – on the ministerial level – and the meeting was held on Monday, 8 October 1973, Shaikh Ahmad Zaki Yamani headed the OPEC delegation, representing the wealthiest oil state, and George Percy (Exxon or 'Esso' Chairman) headed the companies' delegation, representing the largest oil company. In the light of the talks and preliminary contacts, the companies agreed to the principle of amending the agreement and increasing the price, and the bargain was about the extent of such an increase. The OPEC delegation began by asking for double the posted price (from $3 per barrel for light oil to $6). The companies replied that the increase could not exceed 15%. But this is the way with negotiations, where each side begins with the highest level to start the bargaining. And the bargaining did start with Yamani asking for the price to go up to $5 per barrel, that is, the rise of the posted price should be 66% instead of 100%. The companies' team raised the suggestions to 25% instead of 15%, and the situation was frozen at that figure.

The negotiation team held another session on Tuesday evening and it was agreed to hold the third session by the end of the week. But this session was never held, as the companies refused to make any new concessions, asking for two weeks for consultations with their governments. During that time, discussions inside OPEC did not stop, and as usual, there were those who hesitated in cutting the rope, and those who called for a unilateral decision. The situation was speeding up step by step, and by virtue of constant effort, everyone came together to face the question that had been postponed at the beginning but now could not be ignored: What was to be done before the companies' refusal of the demands, when everybody agreed to their validity?

In the face of such a challenge, and in the midst of exciting news about the October War, the member countries were paralysed with indecision. The six Gulf states declared on Friday evening that they would hold another meeting in four days "to take a decision about the method of joint action to define the real value of the crude oil which they produce".

On October 16, the historic decisions of the Kuwait meeting were issued. The producing countries, by their will alone, carried out what the companies had previously refused. The average price per barrel, with effect from 16 October, rose from$3to$5.11, an increase of 70%.

These are the real facts of the story as they happened. And this is the explanation of the sentence in the statement made by the Ministry of Oil and Minerals: "Iraq had played a very important part in achieving this stage".

It was naturally understood that the decision of the Ministerial Committee to determine prices should not affect the rights of member states of OPEC to adjust oil prices according to the clauses of the two Geneva agreements about compensations to producing countries, for losses incurred by them by dollar devaluation. The decision of the Ministerial Committee secured even the adjustment of posted prices of Gulf oil in the future, in the light of the market situation, so that posted prices would rise with the price rise resulting from actual deals in the oil market.

As we may remember, another decision was issued in Kuwait on 17 October, and independently, in a meeting by the Arab oil-producing countries. The decision was to decrease production of Arab oil by certain rates as a means of using oil as a weapon in the war against the Zionist enemy. Undoubtedly, the mere acceptance of the idea by Arab oil-producing countries that oil had a political aspect, and that it was a weapon that must be used in the national battles was considered a development for some parties. The statements, prior to the October War, made by Yamani and Al-Atiqi, then the Kuwaiti Oil Minister, and others, openly attacked this concept. When the Iraqi Command submitted a memorandum about the use of oil in the battle to the meeting of the Arab Defence and Foreign Ministers in Cairo, before the October War, the memorandum was refused by many members, and it was considered a radical proposal.

The important aspect is that the second Kuwait meeting on October 17 was assigned to study and decide the method of using oil in the battle. This meeting was not under the auspices of OPEC, since the members of this organization had no political commitment of this sort. In the meeting, Iraq opposed the method of discussion, and not the aim. Iraq saw that the blow aimed by the oil weapon should be aimed, in principle, at the enemy; friends would have to be protected against any stray measures. Also, neutral parties would deserve special treatment. Therefore, Iraq called for the nationalization of U.S. and antagonistic countries' shares in the oil companies operating in the Arab countries. (Iraq actually did this.) Iraq also demanded the implementation of an embargo policy, or oil-export reductions, according to the political attitude of the state concerned. But this suggestion was not adopted and controversy flared up.

Yet the Arab boycott caused a growing shortage in oil supplies, and the price had to be affected by that, so the companies called for a meeting with OPEC in Vienna. This time, the meeting was not for negotiations, since the stage of price determination by negotiations had ended. The meeting was for dicussion. The Iranian Oil Minister, Dr Amozijar, said in that meeting that the posted price had fallen very low. After this meeting with the companies, the OPEC members decided to hold their own conference in Vienna too, where they announced that the Economic Committee had been asked to prepare a report about the development of prices, to be submitted in the next conference meeting in December, 1973.

As has been explained, the October decisions left the door open for a future price increase when needed. But the Iraqi Command, which led the battle for a price increase before October, expressed reservations and opposition to hasty price increases after October, and made use of the Arab boycott this time. Saddam Hussein declared that Iraq did not think it was a good idea to continually make new and consecutive leaps in prices. The Iraqi officials advised that any new raises had better be gradual. But on December 22, the oil ministers of the six Gulf States met in Tehran. On the following morning, the Shah held a press interview, before the official end of the OPEC ministers' meeting, and stated that the average posted price had gone to $11.65, that is to say to more than double the price decided in October. This raise brought the government revenue to $7 per barrel; in 1970 the figure was $0.93, after the Tehran Agreement it became $1.26, then it went up to $1.76 in September and then to $3.04 in October.

The Iraqi Command once more confirmed the Iraqi attitude towards this raise, after it happened, in the words of President Al-Bakr in his letter to the ex-U.S. President Richard Nixon in January, 1974. The Iraqi President said that the Iraqi Command sees that,

> "the real value of oil – that diminishing element – is much more than its present prices. This is clear in comparing the oil prices with costs of alternative sources of energy used for the same purposes. Then these prices must be understood in the light of the great value of this commodity in the future, not considering oil as ordinary fuel for industry as much as a raw material for production of chemicals, those materials which shall have increasing importance for the international economy in the midst of increasing shortage of such materials.
>
> "It is important to note that present levels of oil prices cannot be considered high when compared to the constant price increases of manufactured goods and basic commodities in the life and progress of peoples. Despite the conviction of the Iraqi Government in all this, our Government has a clear and definite opinion on the subject which we have previously declared on more than one occasion. We have criticised the level and the intervals at which the oil price increases were effected, since we believe that there are certain quarters which want to exploit this matter in the manner which will ultimately conflict with the interests of the producing countries, and influence their relations with the consuming countries, a situation which will ultimately be to the benefit of exploiting states and other quarters, which are more capable of controlling international economic affairs, through their huge wealth. These states and quarters are well-known for their political objectives."

Did events prove this opinion correct? .

To answer the question we have to realize that the oil-producing countries remained unable, after this decision, to raise the price by one cent for 21 months, despite the emergence of circumstances which called for a price raise. Undoubtedly, American moves on the direct political level in the Middle Eastern area, on the oil level, and among the basic consuming states, was able to effect a temporary change in the balance of power among the OPEC states on the one hand, and among the block of large consumers and oil companies on the other. Thus it became possible to contain the power of the producing countries in 1973. But it also seems (given the good intentions of some) that the December leap implied a degree of over-confidence in power calculations, not to say market calculations. And, given the bad intentions of some (which is probable), the interantional cartel and the U.S. Government seem to have had a part in this leap, which would have been utactically unjustifiable by the producing countries alone.

In any case, the question of prices was raised again, after a period of price freeze which lasted about two years. As usual, the Iraqi Oil Minister, Tayeh Abdul-Kareem, presented an explanation of his country's attitude in the 45th ministerial OPEC conference in Vienna, September 1975. He suggested an adjustment of oil prices to meet inflation rates. The Iraqi minister said:

> "The figures prepared by the OPEC experts indicate that the inflation rate in 1974 reached 40% while that rate during the nine months previous to the Conference meeting was about 28%. If rates are to be calculated according to data presented by the International Bank and the International Monetary Fund or the Japanese reports, these rates will be 28.6% in 1974."

Tayeh Abdul-Kareem says that in the meeting he explained that rates of inflation exported to OPEC countries are still much higher than the quoted rates. For these rates are estimated on FOB (Freight on Board) prices, and they reflect higher prices inside countries where these commodities come from. But when we add what we pay to transport these commodities and to insure them, and if we take into consideration the great rates of increase in charges for technology and services, then the actual inflation suffered by OPEC countries in their imports becomes much higher than the rates mentioned above. If the OPEC countries had wanted to fully compensate for their losses effected by devaluation of their monetary revenues, during the 21 months of oil price freezing, then such a compensation would have demanded a rise in oil prices of not less than 80%.

Yet, Tayeh Abdul-Kareem finished this report by suggesting that the rise should be between 25 and 30%, then he came down to 20%. The Iraqi Oil Minister stated that, in setting these rates, he was convinced that they represented the lowest limit of rates of inflation imported in 1974, and the figure was this low in consideration of the international economic situation, and the economy of developing countries, and also to prove the good will of the organization while preparing to take part in the International Economic Co-operation Conference.

This is what the Iraqi Oil Minister said. But we add to this that the Iraqi delegation also listened to the attitudes of the member states inside OPEC, since the atmosphere was charged with storms, and pressures and counter pressures. All this became known, and many observers expected a split inside OPEC into two or more blocs. Some were anticipating this split with apprehension and sorrow, others were anticipating it with spiteful glee.

In fact expectations were not far removed from what was going on inside the meeting halls. In a moment when agreement between different viewpoints seemed impossible, the Iranian delegation – among those who demanded a rise in price – threw in a suggestion that each state announce the price suitable for her oil. There is no doubt that adopting such a principle would threaten the future of the organization. The suggestion may not have been serious, but merely way of pressurizing internal bargaining, but it was unanimously refused anyhow. Yet, it remains that the suggestion was actually made, with all its implications of disagreement.

"Most of the delegations reflected the Iraqi viewpoint," said Tayeh Abdul-Kareem, "and the conference almost reached a suitable decision. But the maneouvres of some delegations, in keeping with the nature of their political systems, and their known external connections, aimed to block any increase in prices and wanted to see to the freezing of such prices, even though this may have led to the destruction of the organization."

**NOTES**

1. M. Field, *op. cit.*, p.10, and ch.2; Sampson, *op. cit.*, chs. 11 and 12; also Salah Muntasir, *The First Oil War* (Cairo, 1975) p.44 (Arabic).
2. Tayih Abdul-Kareem, conversation with the author.

* Information comes from official documents and private conversations.

## CHAPTER 17

# A STRATEGY AGAINST CONTAINMENT

The year 1959 witnessed a general drop in the posted oil price, and the decision was made unilaterally by the companies. In August 1960, the companies decided another slide in the price.

In 1960 also, the desire of the producing countries to oppose the arbitrary decrease in their revenues produced an oil storm with permanent consequences. This led to a conference in Baghdad on the invitation of the Iraqi Oil Minister, headed by Abdulla Al-Turaiqi, the Saudi Oil Minister, and the Organization of Petroleum Exporting Countries (OPEC) was established on September 10, 1960. The Baghdad conference was considered the first OPEC conference, and the oil ministers present were from: Iran, Iraq, Saudi Arabia, Kuwait and Venezuela. Other members joined the organization later from Qatar (1961), Libya (1962), Indonesia (1962), Abu Dhabi (1967), Algeria (1969), Nigeria (1971), Equador (1973) and Gabon (associate member, 1974).

It has long been proved that the emergence of OPEC was a significant event. But in September 1970, there were few who could realize how significant it was. As for the oil companies, it seemed that they gave it a sarcastic smile, and decided, very simply, to ignore the entire affair, declaring on every occasion that they would continue negotiating and dealing with every state separately, and that they would not deal with an organization joining all these states, under any condition.

The first resolution of the first OPEC conference declared that the organization was set up to face the companies, and that,

> "the members can no longer take a careless attitude towards the course followed by the oil companies in effecting price alterations. The members will demand that the oil companies preserve their prices fixed and free from unnecessary changes. The members shall use all means available to them to return prices to the level that was prevailing before the reductions."

Despite the continuous disregard of OPEC by the companies, who always referred to it as the "so-called OPEC", the producing countries did in fact achieve an important part of their first resolution. These countries succeeded in stopping the companies from playing the game of price reductions again. If these countries could not fulfil the second part of the resolution – the sizeable increase of oil prices – then the mere freezing of prices was an achievement, which meant that the companies were no longer free to decide the price unilaterally in practice, though in theory they remained in possession of such power.

The situation remained like this until Libya found it could exploit the geographic features of its oil, which were that it was close to European markets, especially after the closure of the Suez Canal, and formed 30% of European oil imports in 1970). Libya also exploited the variety of concessionary companies operating on her territory, and the existence of a number of independent companies which had no alternative resources in the eventuality of their coming under pressure. Libya used these favourable factors and OPEC moved to a position of attack. In 1971 the Arab Gulf oil-producing countries entered into a battle with the companies, which was a significant turning point in form and content. For the first time, OPEC demanded the right to negotiate as an organization, and issued a warning. And for the first time the companies gave in, and one agreement, the Tehran Agreement, was able to join 13 companies and a group of oil-exporting countries in the Gulf in an equal partnership. This was on the formal level, but in fact these producing countries procured higher revenues, and, more important than this, fixing the posted price was no longer decided by the companies, but through negotiations between the companies and the producing countries. There is no doubt that this was a midway link between the previous situation (pricing by the companies alone) and the later situation of October 1973 (pricing by the producing countries alone).

It has been described how the 17-30 July Revolution, led by the ABSP, broke out in the late Sixties, and that through its comprehensive development of the political, economic, and social situation in Iraq the features of an independent oil policy were crystallizing and growing. All this was reflected in the size of the role played by Iraq inside OPEC. While Iraq was at the stage of direct national development, challenging the concessionary companies in the lands reclaimed by Law No. 80, OPEC was at the stage of negotiating prices. In 1972-3, the stage of Iraq's direct confrontation with the cartel companies and the success of nationalization, OPEC was moving to the stage of prices being set by the producing countries alone and the Iraqi role in OPEC at this stage was a leading one. This confirms the interaction between inside and outside elements: every stage of progress inside Iraq was reflected in external relations, and every stage of progress outside reflected profitably on the Iraqi achievements inside the country.

The cartel which controlled oil production, trade and industrailization was exercising this control according to an international oil strategy, connected with top level strategy of the western allies, and the U.S. in particular. This could not be otherwise with a commodity which is so essential to the way of life of the age, where its control must be decisive in the formation of international relations. If the producing countries began to exercise a growing power in deciding the future of this commodity, then the exercise of this new type of control must lead to influential comsequences on all the traditional systems of international relations, and the powers exercising this new type of control must realize the dimensions of these consequences, and provide a suitable and complementary international strategy to protect the new centre of control, and ensure its development.

The formulation and exercise of such a policy is basically the duty of the advanced oil countries, and Iraq in particular, since Iraq is the only country that exercises full sovereignty over its oil potentialities, and is therefore more capable of developing an independent policy, more capable of independent movement, both inside the country and in the international field. This issue has become more urgent after the important developments in the balance of power since the Seventies, between OPEC on the one hand, and the cartel and imperialistic policies on the other.

I believe that the letter from the Iraqi President Ahmad Hassan Al-Bakr to former U.S. President Nixon displayed a new style of framework for the study of the oil issue from the position of the progressive and independent states. The basic principles presented by that January 1974 letter were:

1. The energy issue is inextricably part of the interests of the entire world and not only the interests of the advanced countries. Therefore the discussion of such an issue must not be confined to large consumers of energy and oil alone.

2. The energy issue cannot be discussed in isolation from other materials essential for world stability like corn, wheat, copper, iron and vital industries.

3. A comparison of the increase in oil prices during recent years with the increase in prices of raw materials and manufactured goods in the industrialized countries will show that the increase in oil prices "despite our reservation on the same" (this was in January, 1974) is

only to preserve the purchasing power of the revenue of oil-producing nations.

4. Consequently, the treatment of oil prices is fully connected with the prevalence of an international system based on mutual respect and interests, ensuring the welfare and peace of all humanity.

Presentation of the issue in such a framework established an economic basis for consolidation between oil countries and the group of economically suppressed countries in the joint struggle towards a more just international system. The Iraqi attitudes in all international meetings have been determined according to these principles. But the Iraqi Command realize that the results of conferences are not ensured , in the final analysis, unless there has been a real change in the power balance, efficient use of all the means of pressure which can affect and strengthen this change. Pressure means and objectives, according to the strategy of the Iraqi state, are naturally economic, political, military, and diplomatic means beginning at home and extending to the outside world. Generally they are: the preservation of the political and economic administration to the utmost limit possible; establishing international relations which help the development of comprehensive international systems through which Iraq likes to operate; and, in all cases, the significant role of the oil strategy within the general strategy. In addition to an independent internal oil policy, a country must exercise special activity in the international oil field, exceeding the activity exercised by a similar non-oil country. Oil strategy also gives the state an authority in international relations far beyond the authority exercised by another liberated country of the same size but not producing a strategic commodity like oil.

Oil can be a means of extending and variating alliances, and it can also lead to enmities and battles. Specifically, oil helps to exploit controversies inside western European countries to a large extent, in such a way as to help oil countries and economically suppressed countries. Oil is a means to enforce equal or acceptable relations between liberated oil countries and big consumers in western Europe and Japan, through the national companies. International oil strategy should encourage the deepening of conflict among the western ranks in order to weaken and gradually isolate the international cartel and the U.S. When the liberated country with such a policy is Iraq, then Iraq is expected to invest this potential in the service of national issues alongside the regional Arab issue in order to achieve a vast and independent development. In this struggle, OPEC occupies an important position in the international economic system as a whole.

Coming to the subject of OPEC, we find a number of principles, methods and controls which guide the Iraqi activities inside this important organization.

Firstly, the Iraqi Command can never forget that OPEC is a front representing a variety of economic, social, and political systems. According to these systems and their international connections, the situations vary in the discussion of each issue. It is usually said that the group of states forming OPEC is divided into a team of "hawks" and a team of "doves" – the radicals, and the moderates. There are some countries which lead a middle course between the two.

Secondly, the oil industry has for the producing countries two basic and complementary aspects: producing oil and marketing oil at a fair price. Up to the early Seventies the monopolies controlled these two stages completely. It is clear that the liberation of oil wealth is not achieved by these countries except by the complete cancellation of the companies' role. The hawks find that this is technically possible, but it demands certain political and social conditions. In any case some OPEC countries have made significant progress in that direction, and Iraq has reached the end of the line. But the co-operation of such countries with the rest of the oil producers, who have a desire and a capacity to skirmish with the monopolies, is a basic question in this ferocious war.

The cartel companies move together according to general strategic lines, and endeavour to co-ordinate their attitudes to the utmost, and to limit their differences with independent companies in the face of the producers. Despite the power of this bloc, a state can organize her powers and succeed in breaking through enemy ranks. But to secure continuity and increase gains, it is necessary to go through negotiations and continuous battles on the largest and most unifed front possible. I have said that for the producing countries the two sides of the oil industry (production and marketing at a fair price) are complementary. But this does not mean that they are not distinguishable, and that each aspect cannot be handled separately, to a certain extent. It has been proved that the price issue is necessarily the most urgent for drawing up a unified policy, and it is, ultimately, the easiest point where an attack by producers can be started, as it is most conducive to unanimity among the producers. So, it is possible to start moving with the price issue, and joint success in the price issue leads to solidarity in other battles against monopolies.

This is the second principle in the Iraqi programme with OPEC, and Iraq has made use of those joint battles, irrespective of their size. At the same time, the success of the hard-line policy inside Iraq helped in the gradual acceleration of attitudes against the monopolies, starting with prices. OPEC started and continued as an organization of price solidarity, but the acceleration of solidarity in prices was basically one aspect of the struggle against oil monopolies' control. Consequently, solidarity had to gradually extend to every country fighting a battle against these monopolies. It was also inevitable that the attitudes of mutual support should encompass new fields, like complete nationalization, partial nationalization, or partnership which have been achieved in various areas in the last few years.

The third method of Iraqi influence within OPEC is that of oil diplomacy. The experience of oil diplomacy, in the producing countries, was limited in the past to the art of managing the conflict, and handling controversies between state and concessionary companies. After the OPEC experience, oil diplomacy began to extend also to the handling of controversies among producing countries. The states which introduce new attitudes and endeavour to push the group a step or more forward are expected, more than others, to develop the art of managing the internal conflict while preserving unity.

There is no doubt that gains achieved during the last few years have underlined the fact that it is in the interest of everybody to see the organization continue in existence, and rise above any political differences among the producing countries. This is generally true. But for the progressive states with ambitious strategic objectives, which are more open to pressure and conspiracies, concern to preserve unity becomes more urgent. Iraq has been careful over this principle at all times. In 1971 the conflict with Iran about Shat-al-Arab was acute, yet Iraq did not object to holding OPEC negotiations in Tehran. Dr Sadoon Hammadi, then Oil Minister, went to Tehran and participated in the negotiations with full co-operation. The same thing happened with Saudi Arabia, since no matter what political storms arose they did not affect the continuation of oil relations with Saudi Arabia inside OPEC.

Beside the conscious endeavour to lead the oil conflict away from the snags of political difference, the Iraqi diplomacy to face controversies inside the oil assembly depends on holding to a firm line of trying to develop the OPEC situations in a realistic manner. Iraq does not try to revolutionize OPEC; and the Iraqi Command accepts and suggests compromises whenever possible, in order to preserve unity.

And finally, under the OPEC general umbrella, the Iraqi Command endeavours to give special support to radical attitudes. In addition to political support, Iraq gave technical assistance to the Algerian Government during its battles with the companies. The Iraqi Government sent about 160 oil experts to Algeria in 1969, and the co-operation is still maintained. The same attitude was taken towards Libya in 1970-1. The Iraqi policy establishes some relations with countries which are warmer than the normal temperature inside OPEC. But, at the same time, Iraq avoids polarization inside the organization which threatens its unity – a difficult diplomatic job indeed.

There are also distinguishable groups inside OPEC, defined by their regional situation, and each having its own characteristics, like Venezuela or North Africa. It is the good luck of Iraq to be in the Arab Gulf area, which represents the largest oil-producing bloc which sits on the largest proven reserves in the world. Undoubtedly, therefore, Iraq's active policy along these lines and its attempts to influence the policies of the area, carry a particular importance.

Finally, there is the Arab group represented in OAPEC, now including: Algeria, Libya, Egypt, Syria, Iraq, Kuwait, Bahrain, Abu Dhabi, Qatar, Dubai and Saudi Arabia. This organization aims at developing co-operation among these countries within the various economic activities of the oil industry. The projects in this area have been particularly active in recent years. OAPEC is here mentioned in the context of the Iraqi relations with OPEC because supporting the economic relations among members of the Arab bloc, through the development of projects and joint interests, has a positive effect on this bloc inside OPEC, and consequently on the Iraqi role.

In the previous paragraphs, the words "hawks" and "doves" were used. These words normally give the impression that the conflict inside OPEC is between the radicals, who demand decisive action, and the moderates, who call for slow development. This may suggest that the dispute is a type of intellectual difference between different interpretations of events, or it may suggest that it is a difference over the method rather than the aim, which is a harmful oversimplication, misleading in an understanding of the real nature of controversies inside the organization. Therefore, it is necessary to immediately remove the misunderstanding, caused by the use of these common descriptions, especially at a time when OPEC is under pressures and prey to constant conspiracies.

The OPEC resolutions of October 1973 were the subject of minute study and analysis by the large consumers. We must confess that they were not so seriously handled by the producing countries, and, consequently, we do not have access to data and information to render precise analysis possible. Yet, we can still touch upon some of the basic aspects in the recent oil developments, which cast some light on the nature of the conflicts at this stage inside OPEC.

Western writings usually came to one of two conclusions.

They either exaggerate the international role of OPEC and the potential role of the petro-dollar, or the cartel companies, and say that the companies lost their capacity for independent action, and betrayed their countries and became a poor annexe to the OPEC countries.

Or, they may reach another, not less exaggerated result, in the other direction. Collaboration between the OPEC countries and the companies became, according to this viewpoint, a tangible fact in the international oil system since the seventies. This collaboration was achieved with the positive encouragement of the US, in what some quarters describe as the "Unholy Pact".

The fact is that the present situation is more complex than that, and both conclusions represent an aspect of the complex situation. In order to achieve a more precise and comprehensive picture, we may say that any observation of the oil situation, since the beginning of the Seventies, may include a number of indisputable facts and developments:

1. The constant struggle of the OPEC countries to impose consecutive changes in the concessionary structure, even going so far as to complete or partial nationalization.

2. The producing countries made during that period utmost use of their struggle against the cartel, in various ways, from conflict between independent companies and the Seven Sisters, to conflict between the large consumers (western Europe and Japan) and the strategy of American control.

3. The achievement of determination of prices by producers alone. Some western governments may decide that oil prices should be high but it now depends on Baghdad, Tehran or Riyadh. Extracting the right to determine the oil prices means that the western destiny will be in the hands of the producing countries to a large extent. This is a real revolution in international relations. It is inconceivable that the US should voluntarily hand over this important right to a group of states traditionally oppressed. The right to issue such a decision is a question

of principle, enforced over the years by the producing nations, which cannot be interfered with or conspired against.

4. The OPEC countries – through using this right to determine the price – were able to get a fair price for their oil in 1973, which was reflected in a boom in their economic situations.

These facts have certain implications concerning the role of OPEC as an organization fighting for the economic interests of producing countries, and against the control of the cartel companies. But, on the other hand, there are other facts which lead to different conclusions:

1. The US used to possess cheap energy, which distinguished her from competitive western industrial nations since the beginning of the century. After 1959, the US lost this feature when low cost oil supplies became available to western European countries and Japan. It is said that this cheap energy was effective in the industrial boom of those countries after the Second World War, even more than the Marshall Plan. It is also said that this low cost oil was one reason behind the low rate of growth in the US, in comparison with the western European countries.[1] As is known, the US policy remained, for various reasons, eager to preserve self-sufficiency in the field of energy, and the oil imports were controlled by share limitations.

In all cases, there is no doubt that the large increase in oil price, especially the leap of December 1973, in addition to random decrease of exports during the Arab oil embargo, put the Japanese and western European economies up against sudden difficulties, which consequently strengthened the competitive position of the US in the western economic system. Naturally the US was certain that the large companies which she controls would undertake to convey these cost increases, or any other increases, to the energy consumers in western Europe and Japan. And, at the same time, these companies would double their profits.

2. This takes us to a further digression about the profit rates. The price increase in a country like Iraq, producing and exporting oil without mediation of companies, became wholly an increase of revenue for the Iraqi treasury. But in other countries – the majority – where the companies exercise control, especially the cartel sisters, these companies achieved unprecedented profits as a result of the price increase. Exxon Company announced that their profits in the last quarter of 1973 rose by 80% in comparison with the same period of the previous year. Gulf Company declared a profit increase of 91%[2]. Chase Manhattan Bank announced in October 1975 that the oil companies

which deal with that Bank recorded an increase in the total revenue deposited in the current account in 1974 of 87%, thus bringing the account to 239.5 billion dollars[3].

There is no doubt that this flow of money into the US, in addition to surplus money from OPEC countries, had a positive effect on the US balance of payments, which would help to cover any increase in the cost of the oil that the US may import. These revenues would also help the US companies to invest in projects to increase energy production inside the US to reach, or try to reach, self-sufficiency in 1980.

3. It is also certain that ex-President Nixon asked in his letter to the Congress in April 1973, for the approval of joint action with big oil consumers to reach a unified policy in the energy issue. It was the US which called for the Big 17 Conference in February, 1974, to insure this co-ordination. Then the International Energy Commission was formed – Kissinger's idea – where all big consumers, except France, participated. The major aim Kissinger had in forming that Commission was to face up to OPEC. His assistant Thomas Andrews said frankly in April 1975, that Kissinger's aim was to destroy OPEC. In September 1974, President Ford raised his voice to say that "sovereign nations cannot have prices dictated to them, or have their destinies defined by artificial arrangement or by the deforming of prices of international commodities". Then followed an acceleration of American pressure and threats, and news leaked out of Washington of plans to occupy oil resources. Kissinger continued his constant efforts to push the Energy Commission. In February 1975, the Commission members decided to decrease their consumption by 6 million barrels per day. This was to weaken the position of the sellers on the market, in order to start a competition to decrease a prices among producers, leading to the break up of OPEC.

Any analysis of recent developments in the world of oil must explain the relationship between all seven of these points, despite their different implications. I think the place to start the analysis is to realize the fact that there is now a duality in the power that decides the movement of oil offered in international trade. There is conflict and interrelationship between these two powers, which must be reflected in decisions and attitudes. These may sometimes be contradictory, or their results may be employed in more than one direction.

The first four represent a tendency which led to the emergence of a new centre of power. The rise of this centre was at the expense of the traditional power represented by the international cartel. The cartel has desperately and incessantly fought over the years to retain the power, but was losing in this confrontation, one round after the other, on points. The recess in the cartel power is at the same time a loss of a basic tool in the hand of American strategy. And, the weakness of the cartel, which had ensured control over oil supplies, came at a time when the U.S. was depending more and more on foreign oil imports, and from the Middle East in particular. The US could not increase her production from oil wells inside her territory; in fact her production of natural gas went down, according to the studies of the working group on energy made in February 1970, during Nixon's term of office. This led the US to increase imports of oil to 6 million barrels per day, the figure of expected US imports in 1981.[5]

Under these circumstances, it seems that the American policy, in conjunction with the oil cartel, reached the conclusion that it was necessary to introduce changes into the oil industry's structure, and the system of administration. OPEC, or the producing countries, imposed themselves as a reality which could no longer be ignored. Also, the strategy of constant confrontation with states which possess a product of such importance and sensitivity and which have reached a high degree of awareness and self-confidence, would only lead at this stage to the strengthening of unity among the producing countries on the one hand, and, on the other, to handing over leadership inside OPEC to the liberated countries which were more capable of instigating confrontation. In accordance with this re-evaluation, the policy changed from disregard and comprehensive war to an official and real recognition of OPEC. If recognition meant recognizing a duality in oil power, it also meant the employment of whatever strength the monopolies still held, and they were still in a very strong position, in order to check the growth of the new power. In brief, recognition meant nothing but a change from comprehensive confrontation to the policy of containment. Attempts at containment and taming demand various methods, including not only threat and intimidation, but also inducement. Amongst the methods used were:

1. The recognition that it is the right of the producing country to increase her share of the oil cake and thus increase her revenues; and also her right to issue influential decisions is recognized. At the same time, attempts are made through various means to curtail this portion of the cake (revenues and decisions).

2. Blockading and isolation of radical countries.

3. Using Zionist aggression as a means of pressure to impose an American peace in the Middle East, and according to American terms, with all international (pushing out the Soviet Union) or internal ramifications resulting from that action.

4. Blackmailing the oil countries with the insinuation of the use of military force.

5. Creating dissent among OPEC members, by making use of artificial surplus and control of marketing to punish the rebels, while raising export rates in the "decent" countries to destroy the price solidarity which is the basis of OPEC.

6. Isolating OPEC from the economically oppressed countries by connecting the economic interests of the oil countries with those of the advanced western world, especially the U.S., thus preventing OPEC from active involvement in the international struggle to change the balance of power. It was hoped that the prevalence of current uncertainties in this balance would soon weaken the positions of the oil countries themselves.

If this new policy of containment had succeeded, the oil-producing countries' bloc would remain intact, but as a minor partner and assistant to the American control of oil supplies, and consequently to the existing international system – the very system which oppressed Asian, African, and Latin American countries, and also supports the distinguished role of the U.S. in the western alliance, which operates through the Energy Commission and seems impervious to any change in the balance of power.

The right of producing countries to issue decisions would have remained but the success of containment would have meant that this right would become formal to a large extent, and that American influence would be the basis for formulating those decisions, though they might issue from the OPEC headquarters.

According to this analysis, those who spoke of a conspiracy between OPEC and the U.S. and the cartel companies had some justification if they had in mind some OPEC members who submitted in one way or another to containment attempts, members who held a really significant position in the oil world.

And those who exaggerated the OPEC role and its capacity to exercise complete control in the near future had some justification too, if they had been watching the revolutionary tendencies, which represent a strategic line, antagonistic to the cartel conspiracies and containment attempts.

The fact is that the apparent contradiction in the events and attitudes in the oil world around the mid-Seventies reflected these two opposing strategies, and the two powers fighting for two radically opposed interests. The liberating countries aimed to remove the cartel and the U.S. as a final objective, and to impose a full and direct control over oil supplies by producing countries, in order to bring those countries into line with the economically advanced countries as far as bargaining powers go. The other camp, in return, acted to prevent this aim, and to submit the oil countries to the camp of the rich, and to the American strategy, in return for a little "pay-off".

Who shall win?

To answer this question, we must not opt out by saying that the people shall come out victorious and that history does not retrogress. This is true only in a general sense, but it is not necessarily true in each separate case. The people may be defeated in this battle or that, and the movement of history may relapse for a few years, although the general movement of history, along the ages, progresses. The entire matter depends to a large extent on the efficiency of those in command to use whatever cards they hold. Those in command of the oil countries have acted with courage and intelligence, during the confrontation stage, until the enemy was obliged to change to the containment policy, which is, in any case, a form of retreat, despite attempts to belittle the size and results of such a retreat. But the command of the battle in this new stage demands invention in planning and tactics, and a readiness to face containment attempts, all of which makes the job more complicated.

In this stage, there are a number of considerations which must be taken into account to ensure the destruction of containment.

All containment attempts, and the success they have achieved, do not rule out the fact that OPEC still has the basic weapon which originally imposed its position and recognition in the field of national relations: that is the possession of oil resources. As well as that, the organization also has an additional asset represented in the gained experience, self-confidence and power which is powerful in influencing the international balance in the possession of atomic weapons.

Also the OPEC countries, by virtue of the power they possess, are able to exercise a degree of independence from imperialistic policy. Some countries could not have exercised this degree of independent administration, under their backward social and political systems, had it not been for the oil. No doubt, this potential existed in the past, and it remains active in facing the containment policy. The possibility of independent action to preserve interests and develop what has been gained remains a basis for the continuation of OPEC unity, despite outside pressures which find a response in some members inside the organization. In the final analysis, the oil-producing countries know very well that oil is a diminishing wealth, and the remaining few years must be ideally employed. In the last quarter of 1975, the balance of payments in some oil-producing countries showed signs of deficit, which is a warning to others, while some consuming countries ascertained great surplus in the same period. Gains are therefore not always secure.

We also need to exercise constant awareness and alertness of enemy movements, with their unexpected surprises, and constant readiness to take the initiative. It is also necessary to keep the OPEC activities and meetings far from the enemy's sight. For instance, Iraq led a movement calling for the removal of OPEC headquarters from Vienna with all its tourist attractions, and from the very building where the Austrian Texaco branch has its headquarters, with all its sophisticated bugging equipment. This may sound trivial, but in fact it is not so.

Another step we must take is to encourage relations with oppressed countries. In the summit meeting of Algiers in March 1975, some did not hide their critical attitude about involving oil with the Third World policies. But the organization was able to win everybody in this round, at least theoretically. This was important but it could develop into a paper attitude if the entire energy policy of the producing countries could not be actually employed to serve this concept, or if oil money remained invested in the West alone, or if the example of development and economic relations in some OPEC countries would strengthen the more antagonistic relations among the special US relations, as Sampson suggests[6].

Yet another consideration depends on the fact that the conflict between western Europe and Japan on the one hand, and the US and the oil cartel on the other, is an objective conflict that cannot be side-stepped through the Energy Commission or otherwise. If the producing countries had made use of this conflict at an early stage of struggle and experience, then there would be even more of a chance to do so now.

The "frightful" potential, which the *New York Times* talked about once, is still real. If the American oil companies were removed from the Middle East, the European countries would have probably made agreements with the Arab countries, and would have produced the Arab oil directly. This would have meant the loss of 1.5 billion dollars a year to the American economy, and the weakening of the American political influence in Europe. The result would probably have been the downfall of NATO, and the appearance of a situation in which most of the west European countries would have been in similar position to France in her relations with the US, and Europe would have been more independent from Washington, in a manner that cannot be compared to the present situation.

And finally, the most important factor in this strategy to stand up to containment is that the Arab nation should become capable of defeating the American infiltration into the area, by developing a national attitude, strong in facing the Zionist aggression. A way of achieving this is to accelerate the role played by the liberated Arab countries, especially Iraq. Included in this role is the successful continuation of the independent oil policy in Iraqi territory, and progress – after control of exploration, production and marketing – towards refining, transport and processing of oil.

This has been a justification of using the two words "hawks" and "doves". Now I think there is no harm in using the word "hawks". But, to use the word "doves" about the "Trojan Horse" and those who are captives of the containment policy planned in Washington is not acceptable. I apologize for the use of this word to – the doves. This does not deny the eagerness of the "hawks" to co-operate, and continue the attempts to convince those who are influenced by Washington that their real interest lies in the unity of OPEC.

## NOTES

1. P.R. Odell, *op. cit.*, p.42.
2. Sampson, *op. cit.*, p.266.
3. J. Al-Mutair, "Multi-National Oil Companies and the Economic Crisis", *Oil and Development*, I,5. (Arabic).
4. *The Oil Fix: An Investigation into the Control and Costs of Energy* (London, 1974) pp.10-16.
5. I. Sa'dul-Deen, *The Economist*, 1,2 (January-February, 1975) (Arabic).
6. Sampson, *op. cit.*, pp.276-7.

# CHAPTER 18

# STAFFING THE OIL INDUSTRY

I soon came to understand from my reading that oil production was not a regular extraction industry. But by understanding about the complexities of this operation were much less than what I saw and experienced in reality. Oil production is a very modern industry, and the staff working in this industry cannot only understand the operation technically or operate the machinery efficiently. More important than that, an oil staff-member should be an "industrial" person in every sense of the word. This means a full commitment to precision and perfection, a realization of the value of time counted by minutes and fractions of a minute, a constant vigilance and complete avoidance of reliance on others, and above all, self-control and assimilation of team-spirit. The oil production industry, because of the sensitivity of the product, and because this product is dealt with in consecutive and complementary stages, using the most developed equipment, demands a first class industrial staff. I can now understand the dread of producing countries at taking on this industry alone, though I do not approve of this dread or of the political consequences which it entails.

The oil production industry involves a great deal of capital, to pay for drilling, controlling the producing well, carrying crude oil to the degassing stations, pumping through pipelines, the storage tanks and the tankers. These stages need various high skills, and their course is controlled by vast complexes, indicators and screens. The least slip may lead to catastrophe, or at least a great loss. For instance, oil and gases which fill the atmosphere of the work site are inflammable. And you have to realize the results of lighting a cigarette by mistake. So, in order not to leave anything to chance, you have to hand over any matchbox or cigarette-lighter that you may have in your pocket.

Industrial security measures are found in all industries. But in the oil industry these are immeasurably more strict. The chairman of the oil company, and his companions in the key positions, must be on call day and night, expecting any emergency which might demand an immediate measure. The network of preventive measures is complicated, and methods of fire fighting are particularly advanced. In addition to the possibility of automatic, immediate isolation of the fire area, the most recent fire extinguishing equipment is ready. Regulations demand altering the chief of the fire department immediately, then informing the highest administrative official.

In any case, the thought of fire is disliked by men in this industry and they are not to be blamed, since this, despite all controls and extinguishing methods, is an ugly and extremely dangerous possibility. It is enough for us to remember that men who work in this industry have this impending danger always before their eyes, and we should understand that this is why this industry, by the very nature of its product, creates this very distinctive type of "industrial" person: precision, importance of time, non-reliance on others, team-spirit. The country which can call on enough staff specialized in this industry has indeed an important human base for industrial development. This was the situation in Iraq. As it has previously been pointed out, the Command was not afraid of faltering in the various stages of the production process after nationalization. Yet Iraq was faced with two problems in this field:

The first was the size of available staff for the stepping-up of oil development, and the economic development projects in general. Here, Iraq found the need for many times more than the number of skilled hands it had available.

The second was the shortage of a new type of staff which was badly needed after nationalization. The available technical staff in the field of oil was trained in operation and maintenance work only, but there was no staff trained in higher administration, planning, or design. And this is a serious point which can be likened to the case of a symphony orchestra. Modern industry in its inter-relatedness is like high standard music making, where the slightest mistake may be disruptive to the flow of the tune. The symphony orchestra is composed of players each representing a special individual skill. But what would the music be like in the absence of the conductor? What would it be like if the conductor were to ask them to do his job as well?

The Iraqi technicians were like this orchestra. In the early stages of nationalization, enthusiasm made up for any deficiency. But the stability of work in the long run cannot depend on enthusiasm alone. Enthusiasm and all moral incentives can only be fuel for a short period of time to help with the fast training in the principles of scientific administration and planning for this modern industry. Revolutionary logic demands the treatment of this problem in a few years, though under normal circumstances it needs a whole generation.

How did I see the solutions to these two problems after these years of actual practice? I shall not deal with the matter here in the language of figures which is available in many official publications. I shall merely relate some observations and impressions.

I was at the control-room of the tool-pusher, or rather the chief of the drilling-team, near the rig. The site was North Rumaila field. At the time of my arrival, drilling had reached 3290 feet down, and there were only 80 feet left before the team finished work and moved to another site. I asked the man to explain to me the gauges all around the room. "This shows me that drilling did not go off the line; this indicates the bit consumption; this shows the speed of drilling; this helps to follow up the weight of pipes suspended." And he went on with his explanation.

Our friend the tool-pusher was a forty-year old giant. He began his career with a contracting company working with BPC, then went to work in Kuwait. When he heard of the national development projects, he returned to Iraq.

"Others beside he returned, and engineers too... We like to work in oil, especially when it is ours... I have 17 years experience behind me, and no one taught me. I used to observe the tool-pusher and learned the secrets of the work. But this took me a long time. The foreigners did not care to teach us. Even when we learn, we were not given responsibility. For instance, I was assistant tool-pusher in Kuwait. My knowledge was even better than the man at the drill. But I could never be a driller and issue instructions.

"Now things are different. The administration has faith in me, and they gave me the responsibility of the rig. Here, the entire group is under my responsibility. Every day I give the instructions to everybody according to their specialization: those who work in the mechanics, on the pumps, in the mud... all, then I receive regular information from them about the progress of the work. Naturally I move around and follow up the work personally, that is to say I do not only rely on their reports or the indicators on thse boards...

"My responsibility is large, and I directly contact the chief driller, who moves about among the wells, but does not come by regularly, though in case of need I can contact him anywhere by wireless, and there is one in his car... No, things are very different now. There were originally very few Iraqis who were equipped for this type of work. But, training now is very fast, the young learn at the Training Centre in Baghdad, and when they come here I explain to them the practice. It took me 17 years to take responsibility of the rig. The young do not need one fourth of this period nowadays. Therefore the company can operate 12 rigs at the same time now. By the end of 1976 the number will be 16 rigs, because I heard that the company will import 4 new rigs for deep drilling... Indeed most of the drillers nowadays are Iraqis, of course there are some Indians, some Egyptians or other Arab compatriots, because the work is so large as you can see. But training is also very fast. You can see that the number of workmen here is large, larger than the rig actually needs, because the rig itself is a school. The process does not depend only on the driller. The drilling crew has to include a number of skilled labourers, and all these have also to be highly skilled in their specialization, otherwise the best driller will fail."

In fact, all the assistants of that man were in their twenties. One of them spoke to me about the role of mud in the work of the rig and how it is like blood in the human body. He also spoke about the materials which are mixed to make this substance which is called mud (which is naturally not normal mud). I was surpised to know that this youth came from west of Qurna, whose people were never formerly involved in the productive work. Does the reader remember the story of the suspicious inhabitants of Qurna in the fifth chapter?

"Where are the rest now? " I asked.

"Most of the younger ones work with INOC. Naturally most of them do work which demands no high skill. But the mere work is a revolution." He actually recruited two of these youngsters. He himself took a literacy course, and went to the primary school, then to the training Centre in Baghdad.

I asked one of the chief technicians at INOC: "Can you really depend on these new trainees to do work of this sensitivity and importance? "

He replied: "It is indeed a problem which may lead to some loss and obstruction. But the loss is bearable beside the achievement. In fact the training centres are not enough to give one hundred per cent confidence to the new trainees. One or two years of exeprience is necessary. What helps in our contest with time is that we have older experienced staff, whom we squeeze, and who make suicidal efforts, whether by direct participation in the work, or by supervision of new colleagues. I was working at BPC and was called to work at INOC. I took charge of INOC work at Fao during the first stage of national development. We had to check, test and operate new equipment. Most of our staff were new, whether engineers or workmen. Therefore it was natural that my working hours should extend from 7 a.m. to 2 a.m. in most cases. The work was not in the office, but in the field. Incidentally services were not available on a reasonable level at that time.

"In this fashion, our young staff received their training at a non-conventional speed."

In addition to this, the programme of preparing technical and industrial staff is now accompanied by a new development in production relations. The basic change in production relations is controlled by the change in ownership of oil production; but the progressive national state administration of oil production is not merely a political or legal change, it is a revolutionary change which had to extend to all areas including the relations of company administration with the workers and the latters' outlook on the question of production. There is the Joint Works Committee which is composed of three labourers and three representatives of the administration, holding periodical meetings, headed sometimes by a labour representative and at others by a representative of the administration. The Committee goes to the work-sites to study the problems and discuss them with high officials as well as with the most junior member of the staff, then submits recommendations for the development of work and human relations. In addition to this job done by the Committee, and that of the industrial relations department, the efforts made by labour union and Party committees in political and legal awareness campaigns began to show tangible results. I have personally experienced the amount of information gained by the graduates of these courses, where the aim is to see the worker connected to the company with more than the ties of wages and material incentives, and to realize the nature of relation between the development and success of his company and the economic development and the building up of a flourishing country where the working masses live happily.

I hope this description is not taken as a form of exaggeration. I do not mean to say that relations inside this company are ideal, and that it is not possible to excel what has been achieved. This in fact is far from the truth, and I was pleased to find a Party official at the Basra branch of INOC expressing this opinion frankly. The Union Committee displays a sign at the entrance of the company in Rumaila saying: "Duties First, then Rights". Educational courses emphasize this concept, but there will always be those who can only think of personal interests, paying little attention to the public good. The administration does not rule out the legitimacy of personal claims when reasonable. There are numerous rewards given to distinction in work – rises in wages, financial prizes, medals, letters of recognition for production heroes. But this does not mean that a constant demand for increasing wages or services without a similar increase in productivity or a consideration for the economics of the project as a whole is acceptable. Yet the blame should not only be placed on the employees, since one can also blame some bureaucratic elements in the administration that do not follow the new system of relations. Sometimes, with the increase in production there may be some delays in services, especially housing, which is needed by the great numbers coming to the fields. We can imagine the situation in a town like Kirkuk or Basra when there is a sudden influx of scores of thousands of new labourers in the oil and industrial projects, when these towns were not prepared to house them so suddenly. Such problems and congestion cannot be avoided, though they may be aggravated sometimes by poor planning or frozen bureaucracy.

The point is that the new areas are not heavens. It is certainly not true that each labourer in these areas is the expected new man. It is also not true that relations among people have become ideal. But the true thing is that building the new man and the new values is a difficult assignment, and that the Iraqi Command is persistently working on this aspect, realizing that it cannot be separated from the struggle to concentrate the technical skill. Undoubtedly, a tangible degree of success has been achieved. Welcome now to the second problem – that of the technical and administrative staff. I mentioned that the first problem was providing the number of staff necessary for enlargement, and was solved by large scale training and contracts with technical staff from Arab countries and other friendly nations.

The second problem was providing special staff for high administration and planning. This, and the necessity of adapting staff in this line, was a very difficult problem.

A high ranking technician in the Iraqi Company for Oil Operations told me that he had worked with IPC since 1958, and had gone up the grades of responsibility until he became Director-General of Pipelines. "In 1958, all workmen, skilled or unskilled, were Iraqis; but the engineers were very few. With time, we grew in number, and there was no one in the fields beside us; but the Iraqi engineer remained an operation and maintenance engineer only. I was expected to operate the pipeline; when one breaks I run and repair it. The installations we received should go to the museum. I hope you believe me! There is no exaggeration here. I am an engineer, and have worked on the maintenance and operation of these lines, which means that I am not reporting other people's talk, but I am reporting my own experience. When we talk about these installations and equipment with foreign manufacturers in order to ask them to supply us with spare parts, they laugh. Some of these engines were discontinued in 1945 or 1948. We knew these facts before nationalization. But, we had no authority to decide to change this equipment. When the foreign administration decided to introduce any improvement, like installing a new turbine for instance, a great number of directors and contractors poured in, and we did not interfere. We were merely operation and maintenance engineers.... Now we are different. We are now creating this, and you can see for yourself. See what we are doing at K2 West, or K1 Unit 13. Everything is our planning and implementation. Hydraulic, mechanical, and civil works, with some of the operations undertaken by foreigners – who are responsible for imported supplies for the guarantee period specified in the contract – all these operations are fulfilled under our direct supervision. In addition to this, we are the party that negotiates to import supplies which we follow up from the factory to the site, and we endorse the choice.... a large problem as you can see. We have changed our mentality from that of operation and maintenance engineer to that of a designer and planner. This was a huge task, and it was the dream of my life. The foreign engineer beside me was a class-mate at the same college, and was not more efficient than myself. But they had planted in our heads that we were not good enough for the process of designing and planning. Please go personally and see the results yourself. Please do not rely on what I tell you only. It was very difficult for us to reconstruct our minds after years of being told you were not good enough. It was not enough to tell your colleagues to go and do a certain thing. It was necessary to convince, and to convince is difficult. It was imperative to work day and night. We discussed and worked 12 hours a day until some of us cracked up and were taken to the hospital."

(In fact it was he who had cracked up.)

I asked this friend whose words were coming out like bullets:

"And you personally, how did you change? "

"Because I believed it was necessary I should change."

"Did you not hesitate when receiving your new responsibilities? "

"God knows I did not! I have had this dream for ten years, without exaggeration. I know the capacity of our staff. From my connections with Syrians I know the capacity of their staff. We recognize their support and we together ensure the arrival of the oil at the tankers. The problem was that former companies were lenient and weak."

"And maintenance?"

"God knows, it is better than before. You may not believe, and you are free to do so. But the criterion is the number of breaks, emergencies and power or water failures. All this is less. The same old lines working with higher efficiency. You can see that in figures."

And the man spread out sheets of graphs and figures, which confirmed his story.

I asked Chairman of the Board of the Iraqi Company for Oil Operations: "Can we say that production economics improved after nationalization in the sense that production costs went down?

"To begin with, I can confirm that the production cost of one barrel of Kirkuk oil was amongst the lowest anywhere in the world, and it became cheaper still after nationalization. Yet, I do not measure the improvement in our oil economy merely by this figure, since the real improvement comes from a new approach to our plans for oil wealth development. We are now much more aware of the principles of studied production designed to preserve our oil wealth. This attitude is different from that of the monopolies, which were mainly concerned with doubling their profits, even at the expense of threatening this wealth in the long run. That attitude made the companies concentrate on production from the cheapest wells, stopping or abandoning production from some wells, on the assumption that they were not economical, though in fact they are so but less than others. Our outlook is different; we are now producing from different oil-traps of different potentialities, and we endeavour to locate the largest number of possible wells in order to utilize the full reserve. All these considerations make it difficult or wrong to depend merely on cost calculation of each produced barrel in order to assess the improvement in the oil industry economy after nationalization. In any case, we have to point out that present expansion in exploration operations and filed development demands a steadily growing expenditure and large investments. When these amounts are added to the produced barrel, especially with the present rate of international inflation (which raises cost of imported supplies) then the comparative study of costs and indications may become difficult or misleading."

The Chairman of the Board of the Iraqi Company for Oil Operations enjoys a measure of authority that was not within the reach of any Iraqi before nationalization. The centres of responsibility in Baghdad, especially the Follow-up Committee, took the place of the London Office. Before nationalization, the aim was to separate planning from implementation, an aim designed and based on political principles. But after nationalization, the aim was to reverse this state of affairs, that is, narrow the gap between planning and implementation, an aim also based on political principles. Under all circumstances, there will always be a degree of difference between those who specialize basically in planning, and those who specialize basically in implementation. But what is demanded now is to enlarge the circle of those who take part in planning, and each man who works in implementation must not be a mere receiver, but should be able to respond and also to send his suggestions. His practical responsibility has now assumed a new dimension by taking responsibility for expansion, concluding contracts, and supervision of foreign suppliers and contractors.

We now come to the construction staff, the project builders. These men play an exceedingly important part in development. They do not only represent a certain technical specialization, but they are united by a similar psychological make-up. They are nomads who do not stay long in one place. They come to a chosen site which may be a barren desert, with unbearable temperatures, or a mountainous area covered with snow, or marshes and bushy land. With the best of luck, the new construction may be near a large city. But whatever the location, these men come to open land, carrying in their imagination the type and size of the construction they want to set up. But these people do not leave anything to chance. Therefore they do not set up their encampment before a prior study of every detail. The "blue-prints" they carry can only be put to practice on the site, and become projects of industrial or agricultural production, a school, a hospital... Naturally, the more complex the project the more difficult the job of the construction men. Countries with deformed economic structures, backward industrially, the countries which had their economy conducted from outside the country for a long time, these countries were deprived in the past of special teams of construction engineers, and were used to seeing projects installed and projects operated by outsiders But Iraq had come a long way by the time of nationalization and had a staff able to operate in and maintain the field of oil. In the field of construction, there was a competent staff, though limited, in the civil works area.

With the tendency towards an independent oil industry, and the general industrial development, it became necessary to develop this important sort of construction team: those who set up metal constructions, install mechanical and electrical equipments. There was no-one in Iraq to prepare "blue prints" for this type of work and there was no one to put those prints into practice. The easiest thing in this case was to depend on foreign contractors to undertake a "turnkey" project, ready to operate under their guarantee. But such a decision means a dependence on foreign sources for large and small supplies. The country which settles for this type of relationship with the outside will remain among the economically backward countries, despite any number of modern factories which that country may be able to import.

Naturally, it is necessary in the first stage to depend on foreign contractors, and following the "turnkey" system in building factories is therefore unavoidable. The planned production installations are ambitious, very modern, very complex, and call for large investments. But to reply on this method demands a collateral decision from the first day that this is merely a transitional stage, and it is necessary to make concentrated efforts to prepare a national staff of growing efficiency who participate at a high level from the start.

The Iraqi Command has actually taken these two decisions in the field of oil: the foreign contractors were extensively depended upon, and, at the same time, a company was formed for Consultations, Planning, and Oil Projects. The company started almost from scratch. and with the development of work during the last few years it has become possible to divide the work now among three technical boards inside that company:

1. The Board of Design undertakes preliminary studies of projects. The planning of this department covers civil, mechanical, electrical and chemical works, in addition to tools and instruments. The department also comprises sections for drafting, survey, and computor work, in addition to a section which undertakes preparation of contracts.
2. The Board of Construction undertakes implementation and construction of projects which can be carried out completely by national staff or in collaboration with a foreign party. According to the nature of this work, the department has stores of materials and equipment, directorates of administrative, technical and civil engineering services. Sometimes there are joint activities between this and the previous department in the implementation of projects.

3. The Board of Pre-fabricated Projects undertakes to conclude contracts with foreign quarters, and follows implementation to see that contracts comply to terms and dates specified. This department includes the sub-divisions of projects directorates, and departments of resident engineers on the sites of these projects.

I hope we should always remember that the Iraqi experiment in this field began almost from scratch. Therefore, the basic projects, such as the Strategic Line and Al-Bakr Terminal, are still implemented by the third department, that is, they are pre-fabricated projects. Real advance along these lines is estimated by how much responsibility for each project is given to the first and second departments. This advance cannot be described simply by the word "difficult". Self confidence cannot be achieved merely by learning. Time, experience and struggle are inevitable. Despite all encouragement by the political command, the "foreigner complex" remains deep rooted. There is the ubiquitous illusion of the supremacy of foreign experience in everything. There is the fear of bearing the responsibility of failure of the project. Many technicians prefer using the familiar system which has been tried and tested. Most of the technicians prefer giving the responsibility to a foreign contractor. This minimizes the element of risk, as giving the responsibility to an Iraqi staff, short on experience, undoubtedly involves a certain element of risk.

The problem here cannot be solved by the mere issuing of a decision. It is inevitable to have to fight a real battle between ambitious elements and conventional or hesitant elements, though these latter may have good intentions. It is said that "necessity is the mother of invention". In same countries, the economic development was handicapped by the difficulties of importing modern equipment and technological know-how. Therefore it was inevitable to have self-reliance to the utmost extent, and to cope with resulting difficulties and dangers. In a country like Iraq, the situation is different. The political command made the theoretical decision to encourage growing dependence on national staff in design and construction. This attitude alone encourages and supports the new staff in taking up the challenge. Yet, at the same time, we must realize that the political and financial leverage afforded by the importance of oil enables the Iraqi experiment to avoid the economic and echnological blockades to a large extent. Iraq is in a peculiar position, since large international companies actually compete in offering tenders so that no sensible country could refuse to make use of this favourable "petrohistoric" or "geohistoric" situation. This is fortunate for Iraq, but it is necessary at the same time to double vigilance so that people do not succumb to the temptations of using imported skill as a substitute for the troubles and risks of using Iraqi hands and minds.

These ideas were swarming in my head when I was talking to a number of high ranking technicians in the State Company for Consultations and Planning of Oil Projects. From the numerous details I heard, one could feel that the wheel had actually begun to turn in the midst of all problems, and despite all temptations of easy solutions.

Today, we find that the company Board of Designs, in the case of pre-fabricated projects, undertakes the general study, and limits the preliminary specifications and technical requirements. Then these documents are collected and offered in an international tender. In other types of project, the Board prepares the detailed documents with outside assistance. Sometimes co-operation takes the form of direct contact between the company and institutions specialized in this type of work (some of the Indian institutions for instance). Other times, co-operation takes the form of contracts with experts to work for the company.

Naturally the local designs begin on the simplest level working up to the more complex, but the important point is that it should begin. This was done for instance with the two reservoirs at Rasafa and Karkh. All economic and technical studies were undertaken by Iraqi staff, and the construction work was done by Iraqis too. The building of giant oil reservoirs is by no means an easy job.

We now leave these generalities to meet a distinguished example of the team of men new to Iraq: the construction men.

In the motor-boat which took us from Fao to Al-Bakr Terminal, I met the man who was the resident engineer of the Al-Bakr Terminal project. He was leading the group of engineers from the Consultations and Planning Company which was supervising the large project carried out by Brown Wood Company. Our friend has a B.Sc. in mechanical engineering from the US (1956) and a Ph.D. in thermo-energy from the Soviet Union (1964). Before his supervision job at the Terminal, he was a resident engineer between (1970 and 4) at the Basra Refinery project. He talked to me extensively about the terminal project, the modern supplies, the use of automation and computors in the various stages of operation, and of their obligation to extend pipes carrying oil for more than 50 km off the Iraqi shores, to reach the water depth needed by the 350 thousand ton tankers.

Beside him were some of his younger assistants who lived and slept all that period in the midst of Gulf waters, "leaping about all day long like monkeys", to use the words of one of them, in order to make sure that the construction was sound in every step. I asked this young man about his previous work and he said he had been an engineer with the air force.

"What is the experience you have gained from the deep-water terminal? "

"In the air force too I had to check the safety of the plane before take-off. I used to be nervous until the plane came back and I was sure that I was precise in doing my job. But at Al-Bakr Terminal, my sense of responsibility has doubled, as has my realization of the value of complete precision. But the nervousness I used to experience in the air force cannot be compared to the kind experienced at the project of the Deep Water Terminal. But it is a delicious nervousness, and an incentive to success, and I do not believe I can now live without it. I believe this is my most important gain beside technical experience. Undoubtedly my technical knowledge has doubled during this project."

I asked the resident engineer:

"Was your work more interesting at Basra Refinery than at the Deep Water Terminal? "

"Working at the Terminal project has the romantic aspect of working on the sea, with all its beauty and unusual problems. But as an engineer, I enjoyed and learned more from working in the construction of the Refinery. Technically, the Refinery was a more complex job, including various engineering specialization – chemical, electrical, mechanical and civil. These varied engineering skills are not needed at Al-Bakr Terminal, despite its huge size. And, the Iraqi staff played a larger part in the direct implementation at the Refinery project. All the civil work was done by the Iraqis, and some of the mechanical work too. Incidentally, an agreement was concluded with the same Czech company to enlarge the capacity of the Refinery from 3.5 to 7 million tons a year, the extra output to be directed completely to export. The important thing is that after training in installing the first stage, the Iraqi staff in the enlargement project became able to undertake the larger share of the work directly. The foreign contractor will only import blue-prints and equipment, with a limited number of foreign experts for guidance. I am very proud of this result.

"For all these reasons I am nostalgic for the days on the construction of the Refinery, which reasons may sound subjective, but, who can say there should be no emotional reasons? ! The work of the construction engineer is creative, and my relation to the project becomes like the relation of the father to his son, surrounding him with his care every day, and rejoicing to see him grow before your eyes. The special significance of the Refinery may be because it was my first child. It was the first project I began to carry out, and Al-Bakr Terminal was my second large project. It is difficult to imagine the pain felt by the construction engineer when the work is completed and handed over to someone else. But we soon forget this feeling when we become involved in the creation of a new project."

The man spoke like a genuine construction worker. These words and this psychological make-up is what you hear and feel with every man in this "international" team. Everywhere they have the same feeling towards what they construct and create, but it is also desirable to add to this feeling the depth of awareness of the aim and value of what is constructed. Undoubtedly, the feeling that what is constructed is for the good of the people, and the building of a society with an independent economy is an incentive to double the efforts and deepen the emotion towards what is built up. And these were some of the qualities of my friend the resident engineer. These men, on all the sites, are a living embodiment of the independent oil policy, a policy constantly confirmed and enhanced.

It is understood that the Iraqi state controls the stages of exploration, operation of oil installations and development of production capacity and marketing. And, the confirmation of this independence demands expansion in all the stages, both vertically and horizontally.

Vertical expansion means controlling planning and achievements in previous fields and stages, with the objective of strengthening the gained footholds of economic independence in the field of oil. In fact, Iraq is now conducting the largest compaign of search and exploration to assess the size of its reserve. In accordance with the decision issued in mid-1974, the responsibility of all new exploration operations (outside previous arrangements with French Alph-Iraq, Brazilian Petrobraz, and the Indian Oil and Natural Gas Corporation) is limited to INOC, which now has 11 seismic survey teams, and there is a recommendation to increase the number by two teams a year. Naturally, the possibility of implementing this depends on providing operational staff for information analysis, programming and result finding. INOC also has 12 rigs working on to exploration findings. Yet, the Iraqi Government welcomes the participation of foreign companies in this exploration campaign, but only on contract basis with INOC, for acceptable charges, and provided that this does not entail any joint project or division of product or any similar arrangements. These contractors may enjoy the right of buying certain amounts of oil agreed upon, but at market prices and not at preferential prices.

Concerning the vertical expansion of the stages of operation and development of production capacity, the preparation of staff necessary for these is growing steadily, as the development of production capacity demands complementary planning to enlarge the entire production line without congestion at any stage. By the end of 1976, production capacity was planned to reach 200 million tons, which is a great achievement.

The decision, according to 1972 plans, was to reach a production capacity of 325 million tons a year by 1980. But this decision was taken when the price of a barrel of crude oil was $2.50. After a standstill in the Iraqi oil industry which lasted more than two years before nationalization, and with the change of situation as a result of price increases, the objectives were revised in 1972 to an aim to attain the best balance between production on the one hand, and the need for development revenues, on the other.

The marketing stage, necessary to secure stability and economic independence (discussed in chapter 15) also needed expansion. Here we come to the twin projects: the strategic line and Al-Bakr Terminal. The decision to implement these two projects, in a record period of time, is an example of political insight which endeavours to provide all necessary security to ensure oil independence and protect export possibilities against decisions that may be taken outside the country, thus preventing an important proportion of Iraqi oil from reaching the shores and markets. The projects of the strategic line and the Al-Bakr Terminal were an insurance against disruption of an independent export policy and against external military and political dangers. It was also a security for flexible movement so that marketing administration could increase economic revenue from exporting across the Mediterranean or across the Arab Gulf, according to the market situation.

So much for vertical expansion in the stages already controlled by the State. In the case of horizontal expansion (in the sense of entering into new stages of the oil industry, the Iraqi policy is taking bold steps, in the direction of: refining, a tanker-fleet, petrochemicals and gas usage.

In the field of refining, in addition to enlarging Basra Refinery, there were plans being made during my stay in Iraq in early 1976 to increase refining capacity to 20-25 million tons a year for export. If this target is reached in 1980, Iraq would have achieved a record rate in this field.

Political decisions were made to approach new fields of oil processing, those of petrochemicals and gas. The petrochemical industry deals with oil as a raw material, to be processed into chemical compounds for various uses. When oil goes into the refinery, it is treated by destructive distillation, whence a group of gases come out, known as oil gases. Then a group of volatile liquids follows, used for combustion in substances like benzine, diesel oil and kerosene. The sediment left after this is a heavy mixture giving heavy oil and naphta. The naphta makes 10% of the crude oil weight, and is considered the basis of the modern petrochmical industry, in addition to the natural gas which can be fully used. The organo-chemical compounds which are now produced from oil or gas have reached 10,000 different types. In fact the failure of the conventional chemical industry to meet the normal needs of the market, or to produce new materials compatible with modern usage, has pushed the petrochemical industry to new horizons. This industry has always represented an important aspect of the contemporary technological revolution in the field of materials, and what it has devised in the way of material and usage exceeds anything that was imagined two or three decades ago. Therefore, it is not strange that the rate of growth in this industry was exceedingly high from the beginning of the Second World War, and that it is an industry with a high investment rate.

To realize the profits achieved from this industry we have to point out that Japan consumes about 250 million tons of oil a year. The 10% naphta left of this amount, together with a meagre import of natural gas, forms the basis of the petrochemical industry in Japan, which gives an annual product exceeding $900 million in value, which is more than the entire revenue of Arab oil in 1971, when that production reached about 740 million tons.

When Iraq signed the contract to construct a giant petrochemical complex and a contract to use gas in the fertilizer industry, it was launching a new stage that aimed to use oil in the fullest possible way. The profit margin from oil processing and sales of processed compounds is higher than the revenue of exporting crude oil. In addition to this, using oil as a raw material for processing industries inside Iraq makes the economic future of oil more stable and consistent, because the progress in petrochemical industries knows no limits, while the life of oil as a basic source of energy does not seem to be long. In addition to that, these political decisions transformed the role of oil in Iraqi economic development from a mere financier to the actual basis for developing a powerful industry. This means that the oil sector became even more active as a component of the national economic structure. In the past, the Iraqi economists used to talk about the duality in the Iraqi economy, and about the oil sector as foreign operation, administered from outside the country, without influence on the rest of the economy. They used to talk extensively about the necessity of joining the oil sector to the national economic structure.

In fact it was sufficient only to visit and observe Basra or Kirkuk in the days of the monopolies to realize the significance of this economic situation exaggerating. One would imagine that after about half a century of exploiting the richest oil fields in Kirkuk one would see some sign of health on the face of this town. The desired interaction between the oil sector and the Iraqi economy could have included other than refining, petroleum industries, and providing industrial and agricultural development programmes with cheap energy. It was expected, at least for show purposes, to see Kirkuk full of clean roads, healthy houses, schools and hospitals. But the visitor to Kirkuk was surprised to see that nothing of the sort had happened, and that the company, even in the residential aspect, had nothing to do with its environment. The town was backward in every aspect. Inside the company walls, however, you moved into a totally different world, supplied with all accessories, gardens, cinemas and clubs.

In any case the wheel is now turning in the right direction. If you go to Basra today you will see how the projects of production in the fields became connected with the construction of the refinery, then with its extension, then the fertilizer plant, the iron and steel complex, the aluminium plant, the paper plant, then with the huge project of liquefaction of gas. With this involvement of oil in the whole fabric of industry, the oil sector is gradually incorporated into the mother economy, and a new Basra is created, not only architecturally, but as a town with a modern mentality.

Incidentally, the completion of the various projects using the gases which result from oil production, in addition to the project of liquefaction of gases, will put to an end forever those flames which exist on the oil fields in all producing countries, which signify the burning of wealth estimated in 1974 at 4 billion dollars.

## NOTES

1. *Oil and the World,* XXVIII, October 1975.
2. M. Ajlan, *op. cit.*, pp. 129 and 146. (Arabic).

# CHAPTER 19

# THE IRAQI CINDERELLA AND THE GREATER JIHAD[1]

The oil country is like Cinderella, touched by the oil angel and suddenly moving from the class of the poor to the class of princes and rich people.

And the oil angel does not deceive any country. The sudden change is known to be something like magic, its effect fades after a limited period, after which Cinderella goes back to her original state.

The legendary Cinderella was not taken in by new clothes, carriages or lights. She did not waste the opportunity or the time. When the clock chimed at midnight to announce the end of the magic period she had already won the heart of the prince, and secured the highest position.

Would any oil country be able to keep its present life-style when the clock chimes and the oil is gone? The heart of the prince, which these countries are supposed to win, is technological and economic progress. Are these countries seriously working towards the aim? And how?

The oil countries neighbouring Iraq were taken in by the clothes and carriages and lights. They drank deep of all that, and did in fact raise the standard of living of their peoples, expand their services and utilities, in one way or another, despite the horrible waste of resources, and despite the apparently uneven distribution of these revenues among various classes and groups. Seeing the overflow of money despite all this, and also realizing that the clock "shall chime", those countries belatedly began to set up productive projects, under the guardianship of the multi-national companies, whether in the form of joint projects, or in the form of projects where the giant monopolies control the technological resources and higher administration, or in specializations complementing the branches of these monopolies in other parts of the world when these countries find markets outside the limits decided by the monopolies in their control of international labour and markets. Even in the projects which produce for local consumption, we find that the companies decide the priorities according to types of consumption created in the oil countries. A certain type of chocolate or perfume cannot be produced on the basis of an objective appraisal of what is needed for the development of this country or that, but because some of the "strong" consumers consume these things. Under the motto of "replacing imports", these branches of industry are set up.

The fact is that the type of consumption cultivated in the majority of these oil countries (which is a copy of the consumer pattern of the western industrial countries, though not the ideal type), does not only affect the priorities in the local industry, which shall remain limited in any case, but what is more important, is that it affects the types of imports. Consequently, these oil countries shall remain bound to the states exporting these commodities and shall consume their financial assets in importing them.

All this is not enough to tighten the bond between oil countries and advanced westen countries. Therefore there is the purchase of shares in projects, the purchase of estates in Europe and America, in addition to depositing surplus money in foreign banks and in purchase of government bonds, etc.

These countries are definitely no Cinderellas. When they wake up to a painful reality, when the effect of the magic touch is gone, and when they find they have not achieved the difficult first effort to reach the stage of regular growth, they will find they have not achieved an independent centre of economic activity, and there will be nothing left from the memory of "sweet" days, except a number of millionaires, living abroad, in luxury.

What about Iraq?

It is easy to imagine that the Iraqi Command plans and fights for a different future, which is expressed in the theoretical documents, and practical achievements of the last few years.

> "The ABSP is a revolutionary Socialist party, which considers Socialism a decisive necessity in achieving the liberation of the Arab nation, its unity and modern growth. Therefore, it is at the forefront of the party's mission to spread Socialistic ideas and values, and to embody them in action on all levels, to endeavour to apply them, according to the requirements of each stage, whenever that is found feasible in any part of the Arab homeland, within a framework of unity, and considers the complementary Socialistic Arab system only ascertainable within this State of Unity."[2]

But how can the material basis be laid for the building of socialism in Iraq? How can socialistic production relations be established? And how can oil have an effect on all this?

To answer these questions in the details necessary needs a whole volume. But since we are dealing with oil policy, we have to shed some light on this important aspect. The preservation and development of an independent oil policy necessitates, in our opinion, a general development towards a complementary socialist system. Conversely, the oil policy basically affects the development potential, and the type of socialist relations possible. But the relation of oil to socialist development needs profound and original studies. If we need to act before the theoretical analysis, then what is also needed, during the actual struggle, is to strengthen our concepts and fill up the theoretical gaps. The Iraqi experiment began the actual practice of the radical changes and development and complete control of the oil wealth and this was a basic support for any other plans. Before the full control of oil wealth, all the talk about change and socialistic construction was possible only as a beautiful dream.

In recent times, and after the full nationalization of oil installations, the contribution of the public sector in the GNP jumped to 68.4% in 1974, while it was 12.7% in 1969. The public sector now covers, in addition to oil, 100% of banking and insurance, 94% of foreign trade, 95% of wholesale trade, all railways and roads, transport facilities, airways, ports, refineries, and a major part of local transport, all projects for water supply, electricity, dams and reservoirs, irrigation and drainage projects, stores of commodities, equipments, tools, instruments and crops. In agriculture, sovereignty is exercised over small and medium estates, and the public agricultural sector now represents 10.2% (1974) of total income in agriculture, while it was only 0.05% in 1969.

Through this central control and development of the basic production resources, the revenues set aside for development rose from 113.2 million dinars in 1967 to 210 million dinars in 1973, an increase of 85.8%. In addition to this high allotment of resources for development, there was a higher rate of financial implementation. In 1967 the finances spent from the amount allotted for development was 61.6 million dinars (only 54.4% of the assigned figure). In 1973 the financial implementation was 150.3 million dinars (71.6% of the assigned figure). Thus the actual increase of financial support for development in 1973 was 142%, compared to 1967.[3]

In 1975, the allotment for the development programme, within the general budget, was about 1,100 million dinars. In 1976 the target figure in the development programme was 1,494 million dinars, a clear increase from previous years. What adds to their importance is that the industrial sector (1976) had the priority (709 million dinars), then comes the sector of industry (268 million dinars). The import policy generally tends to favour production commodities, as these represent about 78% of total imports, at the expense of consumption commodities, which are not needed for development programmes.[4]

The Iraqi experiment in the process of development and transformation to socialism exercises an "open economy". I have pointed this out more than once, assuming that the word was supposed to mean making use of the international situation, and of the possibility of dealing with various states and blocs in a way that increases the profit for the national economy. But the Iraqi experiment exercises an open policy, in this definite sense, according to a number of principles which ensure the desired results:

1. Rejection of joint projects with monopolies.
2. In case of the need for additional revenue, the state resorts to security facilities to obtain finances for projects in the public sector.
3. The choice of projects and definition of priorities are done according to the development programme specified by the state, and not according to preferences of foreign financiers.
4. The projects are offered in international tenders. When offers from western companies equal those from institutions in socialist countries, technically and financially, the Socialist offers are preferred, in recognition of the need to strengthen relations with these countries, and in support of the liberation struggle, regionally and nationally. When the western offers are superior by a marked amount these offers are accepted.

There is no doubt that the political and financial power afforded by oil support the Iraqi administration in achieving this type of international economic relationship.

These are the actual practices. But what is the general theory behind these practices?

The necessity of achieving good steady rates in economic development, and the necessity that these rates should be connected with the objective of correcting a deformed economic structure. This means the cancellation of the previously mentioned duality in the structure of the Iraqi economy, which was one of the aspects inherited by the economically oppressed countries generally. This duality means the existence of an island of modern economic practices, the oil industry in the case of Iraq, in the midst of a sea of stagnant and disjointed conventional economic practices. Cancelling duality is done on the one hand by spreading modernization in all the economic sectors, and, what is more important, joining the modern sector to the body of the national economy and severing its organic connection with the economics of the advanced western countries.

According to the theory of economic stages in the west, the strength of the internal structure of a national economy and the degree of its growth are connected with the type of mutual interaction between various sectors of that economy. In the early stages of development, the mutual dependence among various sectors is very limited. Most of the agricultural output is achieved in units which are semi-isolated, adding nothing to, and taking nothing from, industry. The same thing is true with small industries. Their interaction with their environment is limited. This is generally true, but in the oppressed countries the question was not merely that those countries were at a low stage of economic development. They were not like a child, but like a deformed pigmy, which adds to the difficulty of the situation. Therefore, the type of interaction, or the rate of that interaction among the various economic sectors is not primarily due to the backwardness of the production processes as much as it is due to the duality of the economy and the connection of the advanced sector by a secret rope to foreign states.

So, it is necessary to adjust the economic structure by cancelling duality, and this is achieved in the final analysis through processing oil and national industrial development generally, through the modernization of agriculture and the development of piling structures etc. Generally, this is done through independent development with controlled balance of export and import.

The theory also includes the independent development of regular, balanced production and this demands comprehensive central planning of all economic activities, not just of development programmes.

The economic factors which need to be balanced are: the balance between consumption rates (family and societal) and development rates and the balance between real capital supply and power of work available.

Central planning and commitment to plan also demand that the public sector take the lead in the national economy and development operations. This means that the public sector should control key positions and keep a tight organization to help achieve these objectives.

It is necessary also to develop suitable national and international links to complement the national effort to step up development and secure an ideal Iraqi situation with regard to division of labour inside the Arab world and internationally.

But the most important theoretical basis states that since independent development cannot be based on mere economic calculations, economic development is indeed part of comprehensive changes: political, social, administrative, cultural and educational.

Independent economic development cannot ensure more than one part of a comprehensive and complex development process, planned and led by a revolutionary authority, where the masses faithfully participate. Human labour is the basis, and human labour cannot advance in the direction of economic development, with all its force and excellence, except through these channels.

The theoretical documents of the Iraqi command stress this concept, and warn against falling into state capitalism.

> "State capitalism is a deformed shape of socialism. It negates, or counterfeits democratic relations in production, freezes the role of the working class and paralyzes its activity, makes the bureaucratic officials masters of production, controlling its destiny, imposing on the working class and lower levels of officials a new type of dictatorship, not very different in methods and results from the bourgeois dictatorship, exercising in various direct and indirect methods new types of exploitation, not different in result from bourgeois exploitation. In addition to all this, it is a warped way to look at Socialism if it is believed that it only exists to direct the economic aspect of the growth of state and society....What has been achieved up till now in the field of industry in the form of a comparatively wide industrial base may form a basis from which to move to socialism in the future. But most of the measures taken in this field are closer to state capitalism than to Socialism. The participation of the working class and others working in this sector of production – in planning, leadership and implementation – is still, despite relative progress under the revolution, of a limited nature, wanting in political, economic, and technical efficiency. The industrial production, at present, is controlled, almost in every field, by functionaries where Party men and believers in Socialism form an insufficient proportion. We must confess that some Party men working in this field, who are supposed to be the leaders of Socialist change, have not paid enough attention to Socialist culture and practice. Sometimes they have fallen into rightist attitudes of the old system inherited from pre-revolutionary ages. They have not succeeded in building up relations on the basis of democratic centrality with workers in this sector, and the working class, on an

> acceptable level....In fact, this deviation to a type of state capitalism distracts the experiment away from the path to a real Socialist revolution, because state capitalism is almost a revolution in the legal form of ownership, but it does not necessarily extend the revolution to all the other fields, or by the same degree. This deviation negatively affects the economic development, since it discourages the working classes from taking up responsibility. In addition to that, it threatens the entire experiment by making it vulnerable to the enemies. The organized, participating, believing masses are also the first and last defence lines against conspiracies and pressures."[5]

Up to this point – except for our discussion on merging the oil sector into the national economy – the discussion has been of generalities, concerning the revolutionary practice in any other country as much as the Iraqi experiment. Even our discussion of the oil sector had nothing particular about it except the word "oil". Talking about merging the oil sector- in the previous context – is not different from talking about merging the coffee sector, for instance, or phosphate, or the like, in the economies of other countries which have deformed economies.

But oil cannot be in fact a mere word that may by replaced by another, in the development equation, without having a different result. The existence of oil in the development equation means a great deal. This role which oil can play needs a theoretical re-think to help the oil-producing countries understand a reality that was not previously treated by the Socialist studies and experiments, a reality which was also not treated by serious studies in the west about the development of economies. The theoretical re-think needed is not for the pleasure of academics; it is the means by which to use the oil potential ideally.

What is theoretically clear now is that oil is an exhaustible commodity, and that the wealth it brings is temporary, like Cinderella's magic. It is exhaustible in two senses. The first is the natural sense, that is to say, it is of limited quantity, and, with the current high levels of pumping, it really will dry up in a few years. And we can say that it is exhaustible also as a source of easy wealth, even in the case of its existence in the ground, in the sense that technological advance will definitely find substitute sources for energy, as was done with coal, after a while. So the oil will not continue, as raw material or energy source, as a fountain of financial wealth in the present manner.

In all cases, the revenue from selling growing amounts of crude oil is, to a large extent, selling resources belonging to the state. The merging of this revenue into the national income is, to my mind, a kind of national calculation which is misleading and dangerous. Naturally, even with this merger, Cinderella is still far below the standard of permanent members in the Club of the Rich. If we exclude small states and Emirates, and look at states like Saudi Arabia, Iran and Iraq, we find that Saudi Arabia alone reached a high per capita income in 1974, $3,625, noting that only 25% of the national income was spent at home. In Iran the per capita income was $733.30, in Iraq $600.[6] In the states of the Economc Co-operation Organization the per capita income was $4,825.

It is noted that this per capita income in the OPEC countries had scored an unprecedented leap in 1974. In Saudi Arabia the increase from 1973 was 303% in Iran 293%, in Iraq 216%. In any case, this increase of income has taken the oil states to the gates of the Club of the Rich, if not to the level of permanent members, an honorary membership perhaps. But is this a realistic situation? Is it supposed that the oil countries should believe the story that they are states of high average income, and that they have the right to lead the type of life actually exercised by the rich countries (even supposing this is the required standard! )? Here we go back to the question that oil resources are a source of revenue exhaustible in a number of years. It is therefore a revenue gained from selling resources owned by the state. So, can a sensible administration in any project refrain from deducting from the revenue some amounts equal to capital depreciation? And what happens when the administration considers that all the annual revenues form an added value to be distributed as increase in profits or wages? Naturally there will be great rejoicing at the receipt of higher profits and higher wages. But, ten years later, when the resources are fully depreciated and the income stops, the situation will be completely different.

If this were the case, then the national calculations of national income in the oil countries must resort to a type of differentiation between oil and non-oil resources. In the light of this differentiation, it is possible for those who think according to dry economic calculations only to ask for re-categorization of the entire oil revenues, or the oil resources, into financial resources, so that the society lives only on the income from the development and investment of these resources. Those people may say that, without this arrangement, the society will be living at a standard higher than the standard of its real income, at the expense of wasting some of its wealth. And this is a social injustice, even when complete justice is followed in the distribution of oil resources, because, in addition to the development of society in every direction, there is also the historical development to consider. That is to say that the man who lives to a high standard, at the expense of a diminishing wealth, achieves his immediate well-being, at the expense of future generations, who also have a right to this wealth. The future generations do not mean here a few centuries, but two or three decades only, that is, the baby I carry on my shoulder now.

Yet, because the affairs of life and society are not measured and treated by economic calculations only, the treatment of the oil revenues in this rigid manner is impossible to achieve suddenly, after those revenues have been included in the current expenditure for quite some time, creating living standards which are difficult to change for political and human reasons. Yet, as well as being impossible, it is also undesirable and unjust. The present generation has a real right to a part of the oil revenue to improve their standard of living. Despite the previous discussion, though it was largely sound, the present generation has the right to say it had fought and sacrificed for its interests and the interests of the future generations, in order to secure a reasonable price for the oil resources.

The sum total of oil revenues remitted abroad from the producing countries between 1964 and December 1975, the date of price increase, reached 21.5 billion dollars.[7] This exhaustion could have continued on the same scale had it not been for the fight of the present generation. Therefore, despite some deduction from oil revenues for consumption, it remains true that the present generation has done a lot to preserve the national wealth which shall be inherited. On the other hand, the present generation also bears the burden of the first stage of the development revolution, to help the coming generations inherit capital resources to ensure regularity of economic growth.

Deduction of a portion from oil revenues for consumption by the present generation is quite acceptable then, but provided it be understood that oil resources are not basically an income for consumption expenditure. Portions may be deducted as an exception to improve consumption standards of the present generation. These portions must be limited and carefully estimated so as to maintain the policy of directing the growing rate of oil resources to development investments.

Naturally, we should not forget that turning the oil revenues to new investments is, in the final analysis, a godsend to the masses of the oil-producing country. Without the generous financing afforded by oil, the revolutionary state would have had to suffer in order to scrape together savings to start a development revolution. The revolutionary state which inherited a backward and deformed economy could not cross the starting point without dedicating 20-25% of the national income for development. This enforces austerity measures during the initial difficult years before the fruits of development appear, in a nation which is basically suffering from low standards of living. When oil undertakes the responsibility of financing new developments, it brings down the rate of enforced saving which are otherwise severe in the early stages of development.

This is rationally acceptable, but in reality we find that the balance between what is deducted from oil wealth for current expenditure and consumption, and what is left for development remains a matter for political assessment. This is a very difficult question and demands great effort to have the citizens accept deductions. The question of a necessary balance between consumption and development rates, or between the present and the future, proved to be difficult in all the countries which endeavoured to embark upon development. Research on development problems has treated what is called "the aspirations revolution" which is considered one of the major problems facing the oppressed countries in the early stages of adjusting the course they wish to follow. What is meant by the revolution of aspirations is the aspirations of the middle class, and the other classes in general, towards the types and levels of consumption prevalent in the advanced countries. This aspiration is reflected in political pressures which may oblige the command to make concessions on the side of consumption, thus overstepping what is possible according to objective calculations of the development needs. These concessions naturally lead to a corrosion of the revenues which were allotted for development. Naturally, this concession on the part of the command does not solve any problem for the aspirants, except for the moment. With the corrosion of the development resources, the growth rate goes down, and the consumption problems increase once more in a few years. Political tensions come back more acutely, coups d'état and similar troubles begin to follow one another, until a more decisive command comes to power, and begins to act according to sound principles, and according to the fact that consumption problems and legitimate aspirations cannot be drastically solved and met without a degree of self control for a sufficient number of years.

In any case, this revolution of aspirations is not smaller in the oil countries, but larger. The undiscerning citizen, who has suffered a low standard of living for a long time, is surprised by a sudden abundance of money. The precise nature of this money, resources or revenue, is an idea which needs explanation, and the explanation needs time to be perceived. The sudden duplication of riches takes away the sense, and the logic too, and sets loose suppressed or postponed aspirations, so dialogue and explanation become difficult. What makes matters more difficult is to have a sensible command – like the Iraqi Command – surrounded by oil states exercising policies different from the ones which the former accepts as sound and original. In those countries, the middle classes enjoy high standards of imported consumption. Their upper classes lead a life out of reach of western billionaires, with no thought of tomorrow. The revolution of aspirations in a socialist country here is not imported from another continent, but from neighbouring markets, having the same standard of resources. There is no doubt that this represents a serious challenge. On this situation, the Report of the Eighth Congress says that "It is necessary to continue the policy of limiting the harmful consumption phenomena which affect, one way or another, the controls which strengthen economic independence".[8]

The Report also says:

> "Iraq is situated in an area rich in oil wealth. It is one of the important countries in oil production. To the east and south of Iraq there are some of the richest and largest oil-exporting countries in the world. Among the basic battlefields where the Party, the Revolution and the masses are fighting against reactionary systems and the imperialism which backs them with all its political, material, technical and propaganda power the field of development is of particular importance. These systems try to show to our people that to flourish economically is only possible through attachment to imperialism, and following the capitalist course of development. Therefore, the struggle which is going on now between the Party, the Revolution and all the progressive forces on the one hand, and the reactionary systems attached to imperialism on the other, must be decided by proving the sound nature of the anti-imperialistic course and the sound nature of the Socialist course. In addition to effecting Socialist changes in society, this demands the fulfilment of a comprehensive development plan throughout the country, the constant raising of the living standards of the masses and the following of flexible and developed attitudes

to face this challenge and achieve victory in the end."[9]

"The imperialistic forces, the reactionaries of the area and the rightist forces at home will endeavour to exploit the mistakes that occur in the course of Socialistic change and the accompanying difficulties, in order to weaken the faith in the Socialist course. The question of the living standard of the citizens, providing the basic consumption needs and ensuring the correct quality of those needs will play a basic part in the battle between these two contradictory courses.

"Therefore, with the actual, fundamental and decisive insistence on continuing the Socialist course, with the utmost necessity to develop the ideological work of the Socialist way, the spreading of Socialistic culture and its deepening among the masses, to refute the propaganda of the imperialistic, reactionary and rightist quarters, it is necessary to make the calculations of Socialistic change according to a precise balance, taking into consideration the circumstances of the battle in question."[10]

It can be said that the Command in Iraq tried to preserve a precise balance between political and human considerations on the one hand, and the needs of serious economic development on the other. Yet, some Iraqi economists still used to draw attention to the necessity of making more efforts to stop more seepage of oil money into current expenditure. Despite the increase in the general size of revenues allotted for development, Dr Jawad Hashim says:

"If we suppose that half of the oil revenue at least should be directed to the development budget, this means that not less than 400 million dinars had to be taken up by investment in the country for 1973. But, the central government investments (which are directly financed by oil) did not exceed 198 million dinars, which means that oil revenue that was supposed to go into development but did not was double the amount of central government investment, an indication of inadequacy in the investment intake."[11]

The inadequacy of the investment intake capacity of an oil country needs a separate treatment. But this inadequacy, in any case, was not the only factor in the erosion of the investment share in the period under discussion, as Dr Hashim explains again:

"The rate of pure investment, except in depreciation, of amount allotted for replacement and renewal, fell from 13% of the national income in 1969 to 11% in 1973. This means more seepage of oil revenue to non-investment purposes, while there was a growth in the role of government investment – especially after the Socialist decisions – leading to curtailment of the national saving capacity, and consequently the ability to finance investment. The enlargement in current government expenditure by 11% a year had a share in deciding this tendency, thus becoming responsible for the recurrence of the incapacity of public revenue to face the needs of total public expenditure, despite the continuous fall in investment rates." Therefore, the following is expected:

1. Growth of oil revenue at rates faster than before.
2. Other income revenues remain limited in growth.
3. If that is accompanied by growth in current expenditure at high rates, this will take up more and more of the country's savings revenue especially oil revenue.

(But the figures allotted for central governmemt investment for 1975 and 1976 show the awareness in economic policy of the necessity to take these precautions, and the rise in rates of oil revenue directed to investment. If the actual implementation rate is high, the investment average has gone, according to these figures, to more than 20% of national income.)

Yet, throughout the coming years, the balance between consumption and investment rates will remain among the most difficult and critical decisions in an oil state. In the case of Iraq, the problem becomes more difficult owing to the types of procedure in neighbouring oil states, as has been shown. But the Socialist type of distribution checks the acuteness of the problem. The Iraqi experiment can match the type of conspicuous consumption by a type less conspicuous but more just in distribution. In any case, facing this type of problem cannot be solved by issuing just one decision. This is only a part of a continuous ideological and political battle.

The previous discussion was about a question in the oil countries' economy which has a clear theoretical answer: the oil resources must have the major part thereof turned into new investments, which demands two things: limitation of consumption aspirations, and development of the intake capacity of the national economy to involve the size of large investments projected, especially after the implementation of the first point.

But, after facing the first problem, the Command will be faced by a unique challenge in the question of developing the intake capacity. The political and economic policies will be faced for the first time with the problem of a deformed and backward economic structure, yet one which possesses a great surplus of revenue. The traditional problem treated by all development studies concerned the situation in a deformed and backward economic structure, in need of sufficient financial resources. We are not here just faced with an interesting irony. We are in fact faced with a new problem demanding fresh theoretical thinking.

In the traditional revolutionary experiments, it was not expected to see the total supply of revenues, available for investment, exceeding the limits of demand. It was possible to find minor differences or mistakes in data or planning, or inadequacy in implementation. But in the sum total, these experiments used up or exceeded projected investments, though investment rates were not less than 20-25%. It was not possible to increase the average more than this because the increase would mean a fall in the living standard to levels politically and humanly impossible, and economically too, since the fall of the consumption level below a certain limit has a definite effect on the productivity of labour.

But, away from these limitations, the question always was: could the national economy take up rates of investment largely exceeding the previous ones? This question was usually not discussed, because it represented an impossible supposition in states which did not have large financial resources, and did not expect an overflow of extra revenue from outside resources at effective rates. The question was merely academic. It was more profitable, for instance, to have a theoretical discussion about the minimum level that should be invested, and not the maximum level, to secure a reasonable rate of growth.

But it seems now that the question deserves discussion, after it became possible for the oil states to provide large financial resources. It became legitimate to handle the question seriously: what is the highest limit possible for the investment intake capacity of any country's economy?

It may be said that the question is still of a theoretical nature for Iraq, where investment rates projected are in the vicinity of 20% a year. This rate can be used up by revolutionary changes in production relations, and in the political or cultural institutions, through central planning. It may also be said that the Iraqi rate is still far from the problem under discussion: the highest limit possible. This problem may be raised in Saudi Arabia for instance, if that country had seriously thought of using its oil revenue with the highest efficiency possible.

But this objection is not valid, since the issue, though it may not be on today's agenda, represents a possibility under any suitable international developments in the oil world (as happened in 1973), or by taking into consideration the size of the oil reserve, given that a wise policy in the field of consumption is followed. Then, the question can be posed even with the present rate of investment projected. Investing 20-25% of the national income represents a large size of investment in a country like Iraq, in comparison with what the same rate represented in any state of the same population (like Bulgaria or Cuba) during the early days. This, of course, is due to the relatively high income in Iraq. If investing 25% of the Kuwaiti income means investing 100 units, for instance, then investment of the same proportion in Iraq now means investment of 300 units at least. Moreover, Iraq does not only tend to allot a large portion of revenue for investment, since Iraq – by virtue of its ability to participate in international transactions – is also capable of obtaining highly modern technological equipment, at rates unprecedented in previous revolutionary development experiments.

How can the intake capacity of the national economy provide for all this? What is the ceiling of the country's capacity? This question is on the real agenda, even at the present rates of investment. I do not claim that I have answers to these questions. I only claim that these questions deserve discussion.

But there is at least one aspect of the problem on which I can say a word. It is obvious that the situation in Iraq has changed the traditional development equation for a state facing the problems of the first stage of revolutionary development, namely, the relation between the two variables: available capital and labour. The traditional relation is relative scarcity of capital against availability of a labour force. But in Iraq, the relative scarcity is now in the field of labour, in quantity and quality. Unless this problem is solved, all hopes of development, depending on the great potentials afforded by oil, are shattered.

In 1973-4, the labour force in Iraq was estimated at 27% or 28% of the total population, in itself a low rate. This is due to the fact that women had not yet joined productive work on a large scale (though the literacy rate was the same for men and women alike).[12] Therefore, any tendency to enlarge the labour force needs a great effort to recruit women into the productive army. The increase in the labour force should not only mean increasing the present total figure; it may call for a redistribution of the present labour force over various sectors of activity. According to previous estimates, there is a higher proportion of the labour force among the country people, and a lesser one among city people. This may sound strange, since city people form 60% of the population in Iraq now. With the expansion of industry, it is healthy to see such a change occur in the demography with the emigration from the country to the industrial production centres and the service industries. But naturally, this proportion of city dwellers in Iraq does not reflect this tendency. It merely reflects the desire for government jobs and the services available in the cities. which are financed by oil revenue. And, this phenomenon is shown to be less productive when the section of the labour force in the country (with its smaller population) is larger than that in the city (with its larger population). The figures also show that the number of government employees is more than 400,000 people (excluding the armed forces). Undoubtedly, this figure shows that a large section of the workforce are consuming a slice of the national income reaching 200 million dinars a year, though most of them do not represent a productive labour force. Masked unemployment and partial waste in human resources are understandable if not justifiable in countries with a large labour force. But this cannot be understandable in a country suffering from shortage of the labour necessary for the extensive giant development projects.

With the problem of quantity raised by this type of development, which in Iraq is called "explosive development" – though I do not fancy this expression which leads to believe that it is unplanned development, a dangerous impression indeed – there comes the problem of raising the technical efficiency and variation of specialization in the labour force according to the needs anticipated in the plan. This needs expenditure in unprecedented rates in what is now called "human development". The employment of the most modern technological equipment in Iraq at high rates (in comparison with any non-oil country on the same level of economic growth) is supposed to lead to increased returns of economic development. In addition to that, modern large projects are naturally heavy in capital, light in labour. This is supposed to participate in solving the problem of shortage of labour power. But, on the other hand, it aggravates the shortage of technical labour with specialized skills and high efficiency. The human development, in the sense of spending more on improving efficiency by teaching and training – when reasonable in quantity and quality, and answering the real needs – yields more than development by increasing the number of units of equipment. This is acceptable now. But, in countries of limited financial resources, the general capacity of development remains low, whether priority is given to human development or increase in equipment. In other words, the ability to profit by this "discovery" is decided by the limits of the country's capacity for development. In the countries of large resources, the allotment of large amounts for human development becomes a question of life and death. In addition to the importance of this in all cases, increase in this type of development is imperative for Iraq in the attempt to make use of its oil revenue in the employment of modern technological achievements on a large scale. Without human development on a high scale, Iraq will be obliged to lower the rates of using modern technology, thus losing one of the important features afforded by oil, and the application of revolutionary development will remain in the framework of the traditional version, or else the country may have to continue the import of modern equipment and fail to operate them with reasonable efficiency, which would be an unreasonable waste.

There are tangible efforts being made to deal with this basic aspect of the investment intake problem, and the results of those efforts were mentioned in our discussion of production areas and field development. But, it is necessary to have a clear view of the dimensions of this basic issue in the strategy of revolutionary development, showing that the degree of success in this field largely ensures the highest limit possible of the investment intake capacity is utilized. Our previous discussion of the issue was rather economic and technical, where an expert can limit the extent of change needed in the structure of the labour force, or the amounts of increase needed in each branch or sector of the industry, or increase the total figure, or decide on the ideal geographic distribution etc. Such an expert can also turn the previous ideas into figures for employment, for building new schools and training centres, for the preparation of teachers and administrators needed. He can also recommend the settlement of a number of Arab citizens, contract with others from these or other friendly countries... All this is important and needed, but it is the easy part.

Solving the problem of the shortage of labour power includes technical and planning aspects no doubt. But in fact it is not the most important aspect. Even the recommendations in this respect cannot be implemented bureaucratically by issuing a number of decisions. The question needs a radical change in the nature of the man employed. For instance, it is necessary to implant and confirm the value that respect for work – and not for oil – is the cause of the revenue. Confirming this value is not easy in the face of temptations of living in ease, and without effort, on the oil revenue, while all the neighbours are giving in to that "delicious inertia". Stressing the value of work is not easy, but without this value the wasting of this revenue and unreasonable spending cannot be stopped. And, manual labour has to be respected as well as intellectual and office work. A different look towards the woman's work is inevitable... All this and more is a necessity, otherwise all calculations and data on the real and possible labour power will be meaningless.

But changing minds and values cannot be achieved only through the media. The publicity effort must be connected with changes in production and distribution relations, development of planning, in method and scope, and administrative systems, and also with the development of political democracy. This is realized by the Iraqi Command which realizes the necessity of dealing with the phenomenon of fast accumulation of wealth by some groups of people through speculation, the black market, or extensive participation in contracts. Planning must cover all economic fields and activities especially planning consumption with a sound balance compatible with the stage of development of the country.[13] (In fact the development of giant projects leads to a large extent of interconnection in the economy which is necessary for the building up of a national economy. This doubles the importance of early and precise central planning to avoid severe upsets and congestion.) Administrative systems must also be developed by dealing with their untoward growth and raising their standard of efficiency. All this is connected with politics, with the development of authority and strengthening political democracy, which relies on a larger participation by the masses, and a stronger unity between revolutionary forces and parties under the leadership of the ABSP, on the basis of similar political principles.

In short, the problem of labour power cannot be solved, as a human problem, without revolutionizing these humans who form the society with all its relations and institutions. The process of revolutionizing human beings varies with the constant changes in all fields during the struggle. This process propagates the moral energy which prevailed in the battles against imperialism and the monopolies. What is needed to solve the problem of labour power and enlarge the economic investment capacity to the utmost limit is briefly: a comprehensive Socialist revolution, making ideal use of oil potentialities.

Again, I believe that the issue in this context is not easy. The solution of the investment intake capacity development problem, in this context, is a composite solution. In fact, this intricate fight to develop the intake capacity and increase growth potentialities as a result is more difficult than the struggle of the previous stage against monopolies and obvious external enemies. The fight at this stage is a fight with the self, it is the greater *Jihad*.

Some people may ask here: if all this effort is needed then oil and its money did not do a thing. But this is not true. Oil has done, and it can do, a lot. Whoever once imagined that oil money and potentialities could form a magic solution that could put an end to all problems, and make revolutionary development operate by automation, is definitely mistaken.

The first point about oil is that it can effect a revolutionary development without the need for austerity measures in the initial difficult stage. I have said that making a great effort is necessary to control the new equation offered by oil, which means an abundance of finance and consequently a potential for expansion using modern technology. If such an effort is needed to control this equation, then the success in controlling this new situation yields much more than was ever achieved through the traditional equation, based on a scarcity of financial revenues.

Oil is therefore capable of enriching the Socialist experiment, across the whole range of human endeavour, social and technical.

In planning, production, and marketing, controlling the oil was a wonderfully great achievement of the Iraqi Revolution. The management and treatment of subsequent controversies represented the peak in the art of revolutionary practice. The Command which could fulfil all this cannot fail to achieve victory in the more difficult assignment: the job of using oil to profit the scientific and technological revolution, through a comprehensive and progressive Socialistic revolution on a national scale.

Iraq has the potential for becoming the "oil Cinderella". Incidentally, Cinderella won, despite the enmity and conspiracies of the ugly sisters. Iraq too is capable of victory, despite the reactionary and imperialistic high tide all around.

## NOTES

1 The Arabic word *Jihad,* meaning much more than struggle, fight, or holy war, gains its particular connotation from a saying by the prophet Mohammad, who after scoring his first great victory against the enemies of the new faith, said to his followers: "Now we come from the lesser Jihad to the greater Jihad". The sense is that of preparation for an entirely new way of life under the new religion of Islam. (Translator's note)

2. *The Political Report,* Part V. (Arabic).
3. J. Hashim, *Development Planning in Iraq* (1975) p.80.
4. F. Al-Qaisi (Minister of Finance) *Al-Thawra* daily, April 20, 1976 (Arabic).
5. *The Political Report,* Part V. (Arabic).
6. According to the table in M. Field, *op. cit.,* p.52.
7. Memorandum by the Algerian Government to the emergency meeting of the U.N. (Oil, Basic Materials, Development).
8. *The Political Report,* vol.2.
9. *ibid.,* Part V, ch.2.
10. *The Political Report,* vol.2.
11. J. Hashim, "On the Place of Oil in Development Strategy in Iraq." *Oil and Development,* I,6.
12. J. Hashim, *Development Planning in Iraq,* p.10.
13. F. Qaddouri, (Chairman, Economic Affairs Council, RCC), *Al-Thawra* daily, 25 June, 1976.